Appraisal & Evaluation Library

Application Generation Environments Volume

October 1993

London: HMSO

The Government Centre for Information Systems

Appraisal and Evaluation Library
Application Generation Environments Volume

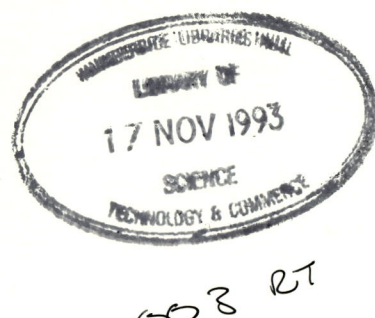

© **Crown Copyright 1993**

Applications for reproduction should be made to HMSO

First published 1993

ISBN 0 11 330604 0

For further information regarding this volume please contact :-

Strategic Programmes Division
CCTA,
Gildengate House,
Upper Green Lane,
Norwich
NR3 1DW

0603 694706

Foreword

This is the Application Generation Environments volume of CCTA's Appraisal and Evaluation Library. This Library is intended to aid appraisal and evaluation of software products and consists of an Overview and Procedures volume, together with supporting technology specific volumes.

The Overview and Procedures volume describes the series and provides a procedure for using the criteria contained in the technology specific volumes in a number of contexts. These include making a strategic selection, evaluation during a feasibility study, and evaluation during the procurement stage of a project. The evaluation procedure is totally compatible with other CCTA recommended procedures, such as those for procurement and evaluation, and methods such as SSADM. It has been written to support the CCTA Information Systems Guides.

Each technology specific volume provides a hierarchy of criteria that may be used as the basis for the evaluation of products in that technology class. Current volumes are for:
- Database Management Systems
- Knowledge Based Systems
- Text-based Information Management Systems
- IT Infrastructure Support Tools
- CASE Tools

as well as this volume on Application Generation Environments.

This Appraisal and Evaluation Library has been produced to assist organisations to identify the product or service, or set of products or services, which best meets their requirements. The procedure and the criteria have developed as technology has changed, and as a result of experience gained from their use. CCTA welcomes comment on, and contributions to, this Library to ensure that it continues to provide maximum benefit.

Appraisal and Evaluation Library
Application Generation Environments Volume

Contents

Chapter			page
	Foreword		3
	Contents		5
i	**Introduction**		9
	i.1	About the Appraisal and Evaluation Library	9
	i.2	About this volume	12
	i.3	About Application Generation Environments (AGE)	15
1	**Application Areas**		23
	1.1	On-line systems	23
	1.2	Batch systems	28
	1.3	Distributed systems	31
2	**Usability**		35
	2.1	Specifying language	35
	2.2	Application development tools	39
	2.3	Generation environment	43
	2.4	Application testing	47
	2.5	Data and application definition	51
3	**Functionality**		53
	3.1	Specification storage	53
	3.2	Specifying language issues	54
	3.3	Components	59
	3.4	Environment type	65
4	**Other system components**		69
	4.1	Capabilities of DBMS	70
	4.2	Integration with DBMS	70
	4.3	Operating system	74
	4.4	Capabilities of the TP monitor	77
	4.5	Integration with transaction processing system	77
	4.6	Data dictionary	80
	4.7	CASE tools	84

5	**Efficiency**		**87**
	5.1	Development productivity	87
	5.2	Development resource usage	88
	5.3	Runtime resource usage	88
6	**SSADM**		**91**
7	**Quality and control**		**95**
	7.1	Documentation	95
	7.2	Development control	95
	7.3	Audit control	96
	7.4	Quality assurance monitoring	97
	7.5	Maintenance	98
	7.6	Performance monitoring and control	99
	7.7	Effect on the organisation	99
8	**Environment independence**		**101**
	8.1	Hardware independence	101
	8.2	Software independence	102
9	**End user interface**		**105**
	9.1	Invocation	105
	9.2	Navigation	106
	9.3	Dialogue	107
	9.4	Presentation and rendition	107
	9.5	Variants	109
	9.6	End user help system	109
	9.7	Session concurrency	109
	9.8	Error messages	110
	9.9	Skill levels	110
10	**Security**		**111**
	10.1	Access control	111
	10.2	Encryption	111
	10.3	Other issues	112

Contents

11		**Vendor and product credibility**	115
	11.1	Quality of product	115
	11.2	Product development status	116
	11.3	Vendor assessment	117
	11.4	Product background	118
	11.5	Documentation	122
	11.6	Training	123
	11.7	Support	124
	11.8	Enhancements	126
	11.9	Related products	127
12		**Project specific requirements**	129
13		**Costs**	131
	13.1	Software	131
	13.2	Hardware	133
	13.3	System operation and maintenance	134
	13.4	People	134
	13.5	Documentation	135

Annex

A	Criteria hierarchy	137
B	Bibliography	145
C	Glossary	147

Appraisal and Evaluation Library
Application Generation Environments Volume

i Introduction

i.1 About the Appraisal and Evaluation Library

This is the technology specific volume on Application Generation Environments (AGE) within the CCTA Appraisal and Evaluation Library. Its subject matter covers the appraisal and evaluation of application development products.

Background

The objective of this Library is to define a framework for

> **'impartial and effective evaluation to find the product, or products, which best meet the needs and constraints of the organisation'.**

The CCTA Information Systems Engineering Division first produced a guide to the Appraisal and Evaluation of Application Generator and Database Management System (DBMS) products in 1986. This volume was updated in 1988 and in the process was divided into two volumes, one for Application Generators and one for Database Management Systems. Both volumes were updated again in 1990 prior to this edition, in which a number of changes have been made including changing the name of the volume to Application Generation Environments.

The Overview and Procedures volume of the Library describes how to appraise and evaluate products and services. Several ways in which the method can be used are also explained.

This volume provides technology specific evaluation criteria appropriate to Application Generation Environments. It should be used with the Overview and Procedures volume.

Audience

The main audience for this volume is Information Systems (IS) staff wishing to carry out appraisals or evaluations for soundly based procurement.

This volume will also be of interest to senior IS management considering the introduction of application development products and wishing to ensure that the process is performed in a professional way, resulting in the selection of the most appropriate product.

It is assumed that the reader has at least a basic understanding of IS/IT, the role of recommended methods such as PRINCE and SSADM and of hardware architecture. Knowledge of application development environments is not assumed, and a brief introduction follows at section i.3.

Because of these assumptions, experienced IS staff may find some of this volume too descriptive in some parts, but technically simplistic in others. The volume may be used as an introduction by those unfamiliar with the topic, and also serve as a useful reference for more experienced practitioners.

Expected uses

It is expected that the volumes in this Library will be used in several ways. The uses identified in the Overview and Procedures Volume are:

- strategic, business-based, evaluation of products to select a strategic product for subsequent organisation-wide use

- less detailed evaluation of products or services as an element of a feasibility study

- full evaluation of products or services during procurement for a project

- independent appraisal of a product or service i.e. not appraised against a specified requirement.

Introduction

Outline of the procedure

The evaluation process comprises seven stages which are described in the Overview and Procedures volume.

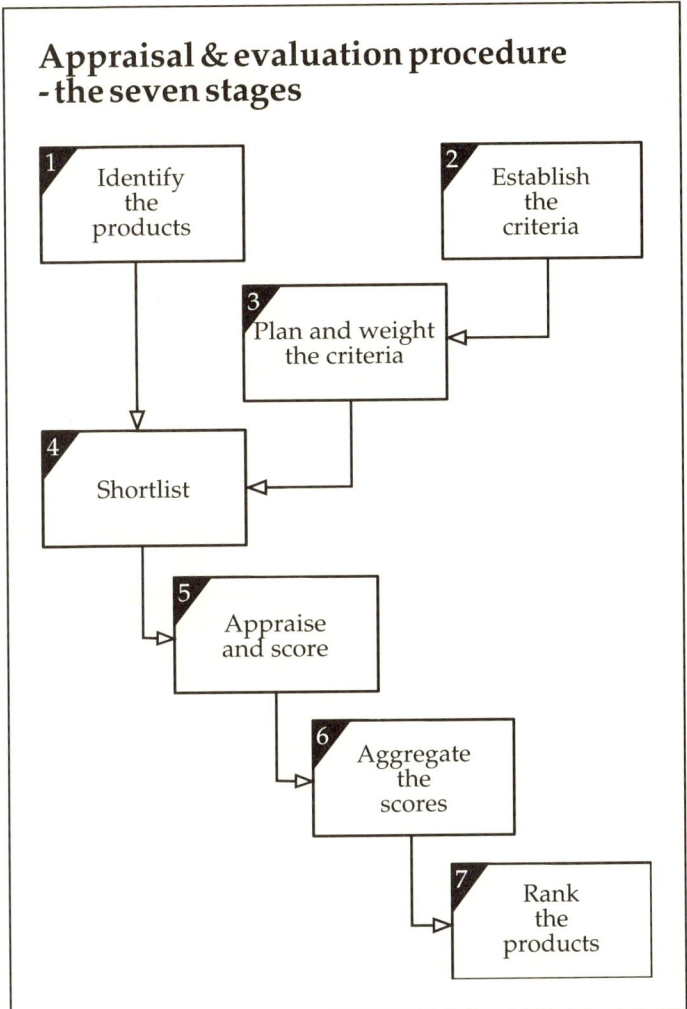

History

This AGE volume supersedes the Application Generators volume of the CCTA Appraisal and Evaluation library originally published in June 1990. The content of this volume has been revised in the light of experience gained through being used in procurements and to reflect the changes in technology and working practices in the last three years.

i.2 About this volume

Scope of the volume
The evaluation criteria in this volume relate to Application Generation Environments (AGE) appropriate for the construction of multi user applications. The method is particularly suited to, and has been used, on both strategic and tactical procurements (See the *Overview and Procedures* volume for further details).

Structure of this volume
This volume is in three parts - introduction, the evaluation criteria and annexes.

The introduction describes the scope of the subject area, the terminology, the notation used for the criteria, and summarises the main headings.

The main part of the volume contains the high level criteria and the checklists of detailed technical questions used within the evaluation model to assess and rank application generation environments. The questions can be used as an aide-memoire when gathering information about products.

The annexes contain a hierarchy chart of the subject matter in this volume, a bibliography and a glossary of terms used . The hierarchy chart may be used as a default or as the basis for a hierarchy chart that is customised to reflect the needs of the project or organisation.

Summary of the criteria
The hierarchy of evaluation criteria against which Application Generation Environments can be scored, is summarised below and elaborated in the chapters that follow. It will, of course, be necessary to construct a hierarchy applicable to the needs of the project or organisation.

Introduction

The top level criteria are:

- Application Areas - the types of applications that can be developed

- Usability - reflects overall productivity anticipated from the product and the skill levels required to use it

- Functionality - the range of functions provided by the product

- Other System Components - capabilities of, and degree of integration between, major data management products which run in the same environment, to provide a total data management capability

- Efficiency - machine resource usage for the development and production running of the applications

- Design Methodologies - extent to which the product integrates with and supports SSADM (Structured Systems Analysis and Design Methodology)

- Quality and Control - of increasing importance for large projects. It applies in particular to the differences between 'tactical' and 'strategic' products

- Environment Independence - important if systems need to be moved to other environments or run on multiple platforms within a network

- End User Interface - ease of use and facilities available to the user

- Security - safety and control of access to the data

- Product Credibility - status of the product and supplier, and degree of support

- Project Specific Criteria - other than above

- Costs.

These criteria, except for costs, should be weighted and scored as set out in the *Overview and Procedures* volume of this library. The cost information will be required as an element of the selection procedure, or to exclude products from detailed consideration when they exceed planned budgets or cost ceilings.

Note that these criteria are not intended to form a tutorial on application generation environments. There is a wide range of published material available.

Questions

This volume consists of a discussion of each of the above criteria together with relevant detailed questions. The questions should be used for familiarisation with a product before attempting to allocate scores against evaluation criteria.

Not all questions are relevant to all products, or projects, and they should be used selectively.

Experience has shown that little will be gained by having the vendor provide written answers to the questions. Only by probing can the evaluation team fully elicit the limits of the capabilities of the products. Maximum value will be obtained by attempting to answer questions after inspection of technical documentation, and attending demonstrations.

Notation

The criteria in this volume are structured as a hierarchy, this is illustrated in Annex A.

Introduction

The text is in three classes:

- the main discussion of the criteria - it is primarily this text that should be customised for particular projects, against which weights are to be assigned and scores allotted. To obtain an overview of the criteria this text can be read in isolation. This is printed in 10 point Palatino typeface (ie the one used to print this volume) alongside a numbered heading in bold type, as illustrated below on this page. Where the criteria covers a large subject area it is divided into sub-criteria. This is printed in the normal typeface with an unnumbered side heading in the same typeface (ie not in bold)

- detailed discussion of the criteria or sub-criteria - this level is required for information gathering. This is also in the 10 point Palatino typeface, it does not have a heading

- *the supporting questions associated with the criteria or sub-criteria - these are in italics, as this example.*

i.3 About Application Generation Environments (AGE)

The importance of Application Generation Environments

This volume is intended to help in the appraisal and evaluation of AGE software. The pace of software development in the last few years has helped to create a systems environment which encourages the use of AGE products. Because this development has taken place in a highly competitive software marketplace, with open procurement policies, a wide choice of products has become available for the application developer.

Data management products are becoming increasingly important; in particular because of their ability to produce high quality, easily maintainable computer systems, faster, and with reduced reliance upon highly skilled technical staff. It does not follow however, that the need for thorough systems analysis and design can be dispensed with. Application Generation Environments (AGE) are used to develop the applications, and Database Management System (DBMS) are used to implement and maintain the data content.

AGE and DBMS products can be obtained from many suppliers, and products from different suppliers can be used together. Therefore separate Application Generation Environment and Database Management Systems volumes have been produced. This allows the separate, independent, evaluation of both types of products. The separate assessment, of the DBMS and AGE components of the software environment, reflects the movement of many vendors towards providing 'open systems'. These permit the various components of an environment to be produced by different software vendors. Other elements such as end user Query Facilities and Report Writers are also likely to exist within the environment but are likely to be of lesser importance.

AGE definition

The 'Data Management' arena can be very confusing because of the multiplicity and contradictory nature of the terminology used by the product vendors and the press. It is impossible to give a universally acceptable definition of terms such as '4GL' and '4GE'. New terms are invented frequently and new products are developed that transcend existing demarcations.

The term 'Application Generation Environment' (AGE) is used to refer to products that are capable of producing a wide variety of applications. The prime requirement of an AGE is the ability to develop applications in significantly less time and with significantly less comprehensive skill levels than by use of conventional programming.

Introduction

Below is set out, in a simple fashion, some of the definitions used in this volume. The fundamental distinction made is between the front-end application oriented elements and the back-end data management elements, with some form of data control or query language providing cohesion and integration. There is no necessity for the front and back-ends to come from the same vendor, indeed many vendors do not supply both.

Front-end

Typically the front-end will include several different elements which are used for developing applications. Some of these will also be used by the application when it is running. These elements may be grouped together and sold as a package, most usually being collectively referred to as a Fourth Generation Language (4GL). This is technically incorrect. The 4GL is a single component that forms part of the required product set. This set is variously described as a Fourth Generation Environment (4GE), a Fourth Generation System (4GS) or, as is used in this volume, an Application Generation Environment (AGE). The term Application Development Environment (ADE) is also sometimes used in this respect but, again, this is an error. An ADE more properly is a superset of an AGE, in that the former also contains CASE tools and other products.

An increasing number of vendors are providing front-end product sets which operate under the mantle of a Graphical User Interface (GUI). This is designed in such a way that all elements have a common look and feel. Specifically, this is why it is now more appropriate to talk about AG Environments rather than just AGs.

Typical front-end tool elements are:

- forms - the ability to design and construct screen images of forms (screens) which can be used for:

- data entry into the database, and subsequent update, query, or deletion

- help information

- menu hierarchies.

In addition, the forms facility often allows the user to create and modify the database structure (often referred to as the Schema)

- Interactive Query Language - to enter data control language commands directly (see Query Facility below)

- Fourth Generation Language (4GL) - usually a very high level language. Examples range from systems enabling complex form and report definition, menu control etc., from within the 4GL; down to products containing little more than commands to control flow through the processing code, and the ability to embed Query Language commands and access to forms, menus, etc

- 3GL Pre-Compilers - to enable query language commands, and sometimes forms, menus, and 4GL modules to be embedded in a third generation language such as COBOL, C or Fortran

- Report Writer - to extract data from the database and format it into structured reports. Often can also write data and thus provide batch functionality

- Data Dictionary - is the repository of data gathered during business analysis, and system analysis and specification. This data can be used for automatic system generation or simply be the directory or catalogue recording the database structure. The data dictionary usually uses a DBMS back-end for physical storage of its contents. Tools like Analyst Workbenches provide access to the data dictionary

- Query Facility - to enable end users to formulate their own database queries. The Query Facilities may be:

  - forms based (see above)

  - Query-by-Example

  - a natural language interface

  - some method to aid the user in building a query language statement

- Decision Support Tools - facilities to draw graphs, or use spreadsheet-like calculators, are available in some AGEs.

These tools may either be bundled together or sold separately. This volume does not include detailed criteria for Report Writers, Data Dictionaries, Query facilities, or Decision Support Tools, although they are discussed to a limited degree.

The front end tools are used to develop applications and to interrogate data. The part of any system which manages and supports the data is known colloquially as the back-end. This back-end consists of several elements:

- Database Management System (DBMS)

- Distributed Processing

- Distributed Database.

Other back-end elements are available for tasks such as the bulk loading of data into the database, for importing or exporting data from a database, or for recovering the database after a failure. These may collectively form a set of database administrator utilities.

Ideally, back-end elements are transparent to both application and developer. Only the database administrator (DBA) will need to understand them, though knowledge of the underlying structure is advantageous to the developer, in order to develop an efficient system and tune it effectively.
The front-end need not be on the same hardware as the back-end. Indeed, separation of the front-end application processing from the back-end DBMS processing is an essential aspect of distributed processing.

The data control or query language is used by the front-end elements to instruct the back-end elements. The two main elements of such a language are:

- data definition (DDL) - The DDL enables the database structure of files, indexes etc to be created and subsequently amended

- data manipulation (DML) - The DML allows data to be input, selected for display, updated, or deleted.

The query language will usually also have the ability to specify security and integrity factors that will govern the use of the DBMS.

DBMS architectures

Database Management Systems may be organised in various ways:

- hierarchical

- network

- relational

- object-oriented (OO) and entity-relationship (E-R).

Hierarchical and Network types are the most widely used types on mainframe computers, though many have now added relational capability. Relational

Introduction

DBMSs, or RDBMSs, were originally implemented on mid-range systems but are now implemented on most platforms. OO and E-R DBMS are still something of a rarity. These are all described in more detail in the *Database Management Systems* volume of the *Appraisal and Evaluation Library*.

DBMS architecture is not a major consideration for AGE appraisal, but the range of products with which an AGE may interface may be constrained by the architecture assumed by the AGE designer. The terminology defined below is an attempt to overcome the problem of duplicity of meanings between older and newer database architectures.

Terminology

Products have evolved in different ways. Consequently they have adopted different terminology for features that meet common objectives.

This sub-section describes the terminology adopted, in an endeavour to find a generally acceptable data management vocabulary.

The objective of an on-line application is to interact with the user in such a way that the user can efficiently carry out his job. The application will accept, process, store, and retrieve data. The user's job might be inputting data, requesting batch processed reports, or obtaining online management information by querying the database. The essence of all these is that the application requests some input, the user types it in, and it is then sent to the computer. We have used the term exchange to refer to this basic 'request, input and send' interaction.

An exchange, therefore, could be any one of the following:

- completion of a data entry form (screen)

- selection of an option from a menu

- cancellation of a help screen

- a single character response to a question, eg 'Accept (Y/N) >'

- sometimes completion of single fields on a form

- direct input of a query language command.

The collection of exchanges into a structure is referred to as the dialogue structure. Typically such a structure will contain menus, forms for data input and query, exchanges that cause reports to be printed or batch operations to be instigated, and often help screens that may be displayed. Further, there will be a convention as to the means by which the user can leave one exchange and go to another or to a menu, that is, generally move about within the dialogue structure provided. This is referred to as navigation.

Processing of the data may be required either before or after an exchange. We have referred to these as pre-map and post-map procedures.

Chapter 1
Application Areas

1 Application Areas

The extent of the AGE's support for Application Areas is a measure of its functional capabilities. In particular, it defines the types of application development for which the AGE is suitable.

1.1 On-line systems

To implement an on-line system the AGE must be capable of constructing dialogues and exchanges. To define an exchange requires the ability to:

- define a non trivial pre-map process (ie processing to be undertaken prior to the display of an input screen). Here, non trivial implies the ability to access multiple record types, or record occurrences, within the process, in addition to normal processing such as arithmetic and conditional logic

- define the application screen format to be displayed and the data fields to be entered. There should be few or no restrictions

- define a non trivial post-map process (ie processing to be undertaken after results have been received from an input screen)

- update or access multiple record types or multiple record occurrences as a result of the exchange

- decide by algorithm, directly or indirectly from the data input, the next exchange to be processed

- allow the user to select the next exchange to be processed

- pass information between exchanges in order to allow the construction of multi exchange transactions.

Implicit within this definition is the concept of an on-line transaction success unit and the associated scope of the success unit.

Practicality

The majority of AGE products may be used to build on-line applications. This is because there is a general trend to develop new systems to operate primarily in an on-line rather than a batch mode. On-line systems tend to be simpler and more effective for the end user, moving control away from the computer department to the user.

Is the AGE intended to be used to build on-line systems?

File handling

Some AGEs place restrictions on the number or the way in which files (or database record types in a database environment) can be handled. For example, an on-line process may allow for the update of one (hierarchically structured) database file only. This would have severe performance implications.

Some systems limit data access for an exchange to one file or record type (or even occurrence). While this is suitable for simple applications, general applications require more general file accessing capabilities.

From one screen (exchange), can the application

- *access more than one record of a particular type?*

- *update, create or delete more than one record of a particular type?*

- *access more than one type of record?*

- *update, create or delete occurrences of records of different types?*

- *access more than one file or type of file?*

Chapter 1
Application Areas

- *update, create or delete occurrences of records within different files or files of different types?*

Screen handling

Some AGEs do not contain screen definition facilities. They either use the operating system's screen definition facilities or they market them as a separate component.

Is the screen definition processor

- *a component of the AGE?*
- *part of the host operating system?*
- *an additional purchasable component?*

Many AGEs restrict screen design. If a new application is being constructed this could be worked round (as the business system design can recognise the constraints imposed by the software and avoid them). However, the restrictions may be a problem in a prototyping environment or where the requirement is to reimplement an existing system.

System defined screen message areas restrict the space available to the application.

Does the application have system generated screen headers, footers or message areas?

Most products restrict the representation of repeated fields to a tabular format.

In what formats can repeating fields be displayed?

General screen handling sophistication can be assessed by checking if the product can support all the facilities provided by terminal hardware (for example, highlighting, suppressing a character display, inverse video, etc).

Can the AGE handle all the attributes you wish to use on the terminals you intend to use?

There may be restrictions on the way in which windows are handled; some products allow the programmer complete flexibility in handling windows. Other products may, however, restrict the format and scope of the window.

How does the AGE handle windows, in particular placement, scrolling and sizing?

Variant screens are particularly useful in providing varying levels of help information on the screen, or to exclude knowledge of certain information from particular users. Using variant screens reduces the need to vary the application code for different users.

Can the screen definitions have variant forms to be used by different categories of users?

(See also chapter 9, 'End User Interface')

Transaction exchange

Some AGEs limit the format or the amount of data that can be passed between business transactions. A temporary data storage area may be necessary to implement multi exchange business transactions. As this may be the only way of implementing a delayed file/database update, a relatively large area may be required.

Can data be passed from one exchange to another? If so, what are the restrictions on its size and format? Is the size and format under the control of the application specification or is it a system default?

Temporary data storage areas should be written to disc so that their contents may be used within system recovery.

Chapter 1
Application Areas

Is the area used to pass data between exchanges, recoverable in the event of a system failure?

Complexity restrictions   There may be significant restrictions in the complexity or nature of the processing statements within a pre-map or post-map procedure. If the AGE does impose significant restrictions, it is important to have the ability to escape into a lower level conventional language, possibly at the expense of usability.

If procedures cannot be directly associated with an exchange then possibly the only way of carrying out input validation, or ensuring referential integrity, may be coding in either a 4GL or a 3GL. If the exchange is used in more than one place consistency may be a problem. Note that many (but by no means all) products place such procedural code in the DBMS to be either triggered automatically or be invoked by the AGE.

Can pre-map and post-map procedures be associated with an exchange at the level of a screen or a field? What are the complexity restrictions on such use?

Help facility   It should be possible to build a help facility for the end users, which can be invoked with a function key, by point-and-click, or some such mechanism, from within an application. It should not be the responsibility of the application designer to have to devise the mechanism for providing this help information. For example, field specific help text could be displayed on demand, or screens of textual information could be defined and chained together.

Can help screens be programmed?

Can help be context sensitive?

How much help can be accessed at all the various levels?

Multi user capability — While usually dependent on the capabilities of the underlying database management and transaction processing systems, some AGEs lack the concurrency control necessary to support multiple applications updating a common database. In particular, some systems on personal and midrange computers claim to provide record level locking of data but their underlying architecture, being based on the host file system, allows only single record occurrences to be locked. In these cases it is possible to construct multi user systems but the design is more difficult and recovery is uncertain.

Is the AGE designed to support multiple concurrent users? How does it achieve this?

1.2 Batch systems

A batch or online system is one which, as a generality, does not require regular and/or frequent responses from the user. The processes may be time consuming or involve complex processing and as such can be run out of normal working hours. The main characteristics of a batch system are that it can:

- be driven from a sequential input file containing 'business transactions'. This file could contain a single 'transaction' that initiates the production of a report. In such a case, the input transaction file would usually be referred to as a control file

- define batch business transaction boundaries (for recovery, integrity and performance purposes). Note however that the entire process may constitute a single 'transaction'

- perform non trivial user defined arithmetic, conditional or control operation sequences

- process (ie be able to read, optionally update, or create) multiple record types and multiple record occurrences within a batch business transaction

Chapter 1
Application Areas

- produce printed output or reports in user specified formats.

Practicality

Several AGEs are not capable of constructing batch systems. This is because AGEs are generally more cost effective when used to build on-line applications. In such environments, purpose built report writer packages may give some support to the batch application requirements.

Is the AGE capable of being used to build batch systems?

Do batch applications run as independent programs or do they execute under some form of AGE runtime control routine?

Use of a procedural language of some kind is usually required in order to permit some form of control within the application.

How can batch applications be implemented?

It is desirable to be able to initiate a batch job to run in the background from within a generated on-line application. Having to use a job control language is not end user friendly.

Can batch applications be initiated:

- *via job control?*

- *from within a generated on-line application?*

- *from a host Transaction Processing system transaction?*

File handling

As for on-line systems, there should be no restrictions or limitations on the way that files or databases can be handled. In particular, the ability to update several files or database record types is essential.

To provide a general batch capability it is necessary to have a general file handling capability.

Can a batch application:

- *access more than one record of a particular type?*

- *update, create or delete more than one occurrence of a record of a particular type?*

- *access more than one type of record?*

- *update, create or delete occurrences of records of different types?*

- *access more than one file or type of file?*

- *update, create or delete occurrences of records within different files or files of different types?*

Reporting

Not all products have comprehensive capabilities for defining report formats; this is analogous to restrictions on screen formats for on-line AGEs.

Does the batch generator:

- *produce file/database update programs?*

- *have report writing capability?*

- *produce output to a file/database for later processing by a report writer?*

Can a batch application produce reports on a printer from databases or files?

Complexity restrictions

Complex logic may need be to be used in developing the control structures and algorithms when building batch programs. The AGE should not impose serious restrictions.

Chapter 1
Application Areas

What control structures can be included in the batch application? Can additional processing, for example 3GL modules, be included?

Multi user capability   As for on-line systems, multi user capability is usually dependent on the facilities provided by the underlying database management system. However, it is necessary to be able to exploit the facilities provided by that software; in particular to be able to define success unit boundaries, and to be able to predefine file or database area usage modes, so that the DBMS can invoke the correct record locking strategies.

The ability to run batch and on-line applications concurrently on the same data probably depends on the features associated with the operating system and DBMS. Note that possibly even on-line applications might not have concurrent data access.

Can batch and on-line applications have concurrent access to the same databases (files)? If so, what facilities are provided to control concurrent access and maintain data integrity?

1.3 Distributed systems

The characteristics of a distributed system essentially combine those of on-line and batch systems (above), within a distributed environment. In particular, this relates to distributed database and/or client/server approaches to distributed systems. The fundamental issue is the fact that any one dialogue may incorporate data that has to be retrieved from multiple locations. This may also apply to the development process as well as the eventuating application.

Practicality   There are multiple methods that enable distributed systems, but they all include networking. Some AGEs provide specific support for certain networking protocols, while others provide their own networking product. Use of an AGE that has no such support will not necessarily be precluded from being able to run, but it will be necessary to ascertain any restrictions that may apply.

Which networking protocols will the AGE run with?

How will the AGE run with your chosen or existing network operating system or protocol?

If the AGE has its own networking product how compatible is it with any existing protocols?

Will this impose any restrictions on the use of the AGE?

File handling	As for on-line and batch systems, there should be no restrictions or limitations on the way that files or databases can be handled. In particular it may be necessary to access data, from a single form (screen), that resides in more than one database or file system.

Is it possible to concurrently access more than one database?

Of course it is conceivable that you might have two distinct databases running on the same hardware within an on-line application. In this case the question might also apply within that environment.

Screen handling	As for on-line systems. This function should be divorced from the distributed nature of the application.
Inter-process communication	This is essentially the same as a Transaction Exchange within an on-line environment. It is possible, however, that there may be additional considerations with regard to exchanging data between processes running on distributed processors.

Does the fact that the AGE is to be used in a distributed environment, impose any additional restrictions on its use with regard to transaction exchange?

Chapter 1
Application Areas

In particular, there may be different databases in different locations. Applications may require access to both. To achieve this it will be necessary to join across the two databases.

Does the AGE support joins across multiple databases?

Data integrity

A fundamental issue for all distributed applications is how to manage the update process. Suppose, for example, that stock is to be transferred from one warehouse to another. Databases at both locations need to be updated. But both updates must be accomplished successfully, otherwise the data integrity of the network will be compromised.

The standard method for accomplishing this, is by implementing a two-phase committal process, by which either both databases are updated or the transaction is rolled back. Normally this is a function of the DBMS, but not all database management systems support a two-phase commit, and some are better implemented than others. In any case the programmer should not have to know about it.

In a distributed environment, does the AGE support database provision of two-phase commit, in a manner that is transparent to the programmer?

Where the DBMS provides two-phase commit, is this XA compliant (that is to the X/Open specification)?

Where the (existing) database does not support two-phase commit, what facilities does the AGE provide in this regard?

Complexity restrictions

Again, these are essentially the same features as for on-line and batch applications, with one exception.

It is sometimes the case that development teams are widely dispersed geographically. In this case, particularly if a distributed database strategy is in place, then there will be data dictionaries in each

33

location. For application development purposes it will be useful if a definition defined in one dictionary can be used by another dictionary that is physically remote from the first. This means that a collection of physical dictionaries can be logically viewed as a single dictionary. Without this capability the development process will be severely encumbered.

Can multiple physical data dictionaries be logically viewed as a single dictionary?

It should be noted that the same two-phase commit process (discussed above under data integrity) will also be required for update of a distributed dictionary.

Help facility   There should be no difference between distributed and on-line systems in this regard.

Multi user capability   Again, there should be no distinction as from on-line or batch systems, whichever is appropriate.

2 Usability

The ease of use of the product, when developing applications, is a major factor in determining the overall productivity level that is potentially offered by the product. This includes the ease of use of the application development tools, the learning curve and whether the tools can be used by end users to build their own enquiries.

2.1 Specifying language

Vendors of AGE products are inclined to advertise the non-procedural nature of their offering. In a language of this type it is, theoretically, not necessary for the developer to define how a problem is to be resolved, but simply requires a statement of what needs to be accomplished.

In practice such systems are difficult to achieve. They are also extremely difficult to debug because it is not a simple matter to discover what the software is doing. For this reason most vendor's products, despite their marketing claims, tend to be a mixture of both approaches.

In particular, most products are event-driven. That is, individual modules of program code are linked by the design of the form (screen). If, for example, an entry is made to a data entry field, this event might trigger a particular module of code before returning control to the form. Thus the overall development would be non-procedural in its forms handling, but procedural in its code modules.

To what extent is the language procedural?

Is the language event-driven?

Where applicable, how do the procedural and event-driven elements of the language interface?

In the last few years object-oriented AGEs have been introduced by a number of vendors, utilising sophisticated Graphical User Interfaces (GUIs). Since these adopt a rather different approach from traditional AGEs, these are considered separately. We commence with the more orthodox approaches.

Specification type

AGEs vary widely in the ways in which applications are specified. Extremes vary from a pure declarative approach, through 'form filling' type languages, to more conventional looking programming languages. These vary considerably in their usability and by whom they are capable of being used.

Form filling systems are more usable for inexperienced users. They are also probably preferable for prototyping.

What type of application specification language, or languages, are available

- *declarative?*
- *form filling?*
- *3GL type?*

Object Orientation

Vendors who offer products based on GUIs typically extol their object-oriented (OO) nature. These environments are invariably of a form-filling nature using point-and-click and drag-and-drop methods. These generally require less programming skill on the part of the developer.

More significant is the extent of the object orientation embedded in the product, since some suppliers are inclined to equate GUI with OO, whereas they are not at all the same thing. It should be noted that an AGE which is fully object oriented is likely to require retraining, in programming techniques, for those not already familiar with the approach.

Is the AGE object oriented throughout?

Is the AGE only object oriented in so far as its manipulation of screen based features are concerned (ie writing of procedural code modules is conventional)?

Does the AGE provide object type features for manipulating screen-based objects?

The essence of the provision of object orientation is support for class structures. This will provide a range of functionality, the most important of which is inheritance.

In a complete OO implementation, products will adopt the concept of inheritance. This means, for example, that a father class can be defined, variants can be defined as children of the father, further sub-variants can be defined as grandchildren, and so forth. Now, inheritance means that any change to a class will automatically be reflected in all its descendants. AGEs with GUI capabilities do not necessarily include such facilities.

Does the AGE provide inheritance? Is this for screen-based objects only?

Suitability for non programmers	A potential advantage of the AGE approach is that by removing the technical skill requirements from constructing applications, they can be constructed by analyst/programmers, thus avoiding the problems associated with the analyst producing program specifications for the programmer to code. While some AGEs are to an extent usable by people with little formal programming training, others require conventional programming skills. DP skills are, however, still necessary to understand the concepts of good system design etc.

What elements of the product can be used by people without DP skills?

What elements of the product can be used by analysts?

Do any elements of the product require a knowledge of programming?

Do any elements of the product require a knowledge of database design?

Application area compatibility

The language, syntax and system development environment for constructing and testing batch systems, should be identical to that for on-line and distributed systems, so minimising programmer training and support overheads.

Is the same language and development environment used to specify batch, distributed and on-line applications?

Maintenance

By far the greatest cost in developing applications of all kinds, rests not so much in the initial development but principally in on-going maintenance. Surveys estimate that this typically represents 80 to 90% of the total cost of a project.

On-going maintenance is of two types. First, there is bug fixing, the requirement for which will very much depend on the quality of the initial development. This will be discussed in Chapter 7. More significant, there are always requirements for add-ons to, and further development of, the initial product. These can also be a further source of errors, not just in their own right, but also in the way in which they impact upon the existing application. This can often occur in ways which are unforeseen by the developer. The modular approach of an event-driven environment may be helpful in this regard, since individual pieces of code are effectively isolated from one another. Nevertheless, the problem may still occur, particularly since

Chapter 2
Usability

individual modules may themselves require modification.

Most of the facilities that are useful for the maintenance function, also apply within the initial development environment and are covered elsewhere in this volume. In particular, these include documentation, object libraries and so forth. Version control should also be included for the development process but is especially significant here.

Does the AGE provide source code version control?

Another facility that will prove useful in maintenance of this type, will be the use of an active or dynamic, as opposed to passive, dictionary. This is discussed in more detail in Chapter 4.

2.2 Application development tools

The usability of tools available to the application developers can significantly affect their productivity.

Specification editor

The minimum requirement is a purpose designed 'full screen' editor or editors for screens, program code, administration details etc, with the movement between them being transparent to the user. If the application takes the form of a series of program like statements, sophisticated editors make the program writing easier, quicker and much less prone to errors. A sophisticated program editor should allow short codes to be used for standard constructs and prompt the user by automatically inserting complete construct skeletons. This allows the system to maintain a neat, indented program layout. It also allows the user to manipulate entire program constructs rather than individual lines or symbols.

Own, purpose built editors are usually the most user friendly. Using host operating system or dictionary editors usually implies switching to another environment to edit an application specification.

How is the application specification maintained:

- *host operating system file editor?*
- *data dictionary editor?*
- *application generator's own editor(s)?*
- *other (please specify)?*

Does the editor automatically provide construct skeletons?

Line and context based editors are the least user friendly. Full screen editors are most suitable for 'expert' mode working. Formatted screens are best for novice users. Note that it is possible that a generator will include more than one editor, or that the user may make his own choice of editor.

Is the editor used to maintain the application specification

- *line based?*
- *context based?*
- *'full screen' based?*
- *based on the use of formatted screens?*

Does the system present a seamless interface to the user?

If not how does the generator provide editing facilities to the user?

Where more than one editor is used, is the movement between editors transparent?

Can the user make his own choice of editor?

Chapter 2
Usability

Some systems completely separate specification maintenance from compilation. Interactive syntax checking helps to correct errors early.

Does the product's application specification editor (or data entry system) include interactive syntax checking?

Screen definition

When a screen has been defined, it is important to be able to edit the definition without unnecessary respecification. Some packages make a poor attempt at screen painting, using marker characters and optional field coordinates.

A significant difficulty with screen painting is the specification of the multitude of attributes and field names required. One way is to split the screen into two windows, with part of the users screen image in one window and a 'fill in the blanks' formatted screen in the other. An alternative is to have pop-up windows that appear as required.

If the screen definition is within the application then it is probably not easy to have common screens in several exchanges. If the host system's screen definition facilities are used then this probably means either having to learn another language or having to move from one development environment to another during an application specification.

How are the application's screens defined

- *interactively, using a 'paint the screen' technique?*

- *by drawing the layout in a text editor?*

- *by specifying field names and coordinates on pre formatted help screens?*

- *by specifying field names and coordinates within the application definition?*

- *by specifying field names and coordinates in the screen definition language of the host system?*

- *any other method?*

How are attributes specified?

Screen painting alone is acceptable in a prototyping environment, but for production work it is necessary to be able to edit the screen definitions for consistency.

If a screen painting technique is used, can the definition be edited?

It is necessary to establish whether a screen definition can exist as a separate application-independent entity, for use in more than one place.

Can one screen definition be used by multiple applications?

When defining a screen, it also should be possible to declare the validation rules, error messages and other pre-map and post-map functionality. Ideally, validation should be held centrally, to be enforced globally. Particular local constraints also may apply. Local error messages are important because differently skilled users may use different parts of the application.

Can validation rules, error messages and calls to other procedures be associated with individual fields or complete screens? How complex can this be?

Graphical user interface	If an AGE is based on use of a GUI then its degree of usability will largely depend on the extent of the features embedded within the product. If, for example, you wish to place a scroll bar within a form, then the ability of the AGE to support drag-and-drop for scroll bars, will be fundamental in considering the ease-of-use of the product. It should be noted, however, that such a capability would be completely outside the

scope of a conventional AGE. Thus we are here considering the relative merits of GUI based products.

With respect to the considerations above, that is, editing and screen definition: these will all be inherent in the graphical nature of the product. However, it should be noted that writing of any procedural elements of an application will simply be a windowed version of a more conventional approach. Thus questions regarding editors remain valid in this regard.

The relevant features of a GUI will be discussed more fully in chapter 3: Functionality.

2.3 Generation environment

It is desirable that the AGE may be used effectively without a significant programmer learning overhead, and that applications can be developed and tested with minimal delays.

Components

Ideally, the user of an AGE should only be required to master a single environment, which has all the facilities he needs.

How many different components does the product have (i.e. screen painter, menu builder etc)?

How many different approaches does the developer need to learn before a complete system can be created, ie is a screen developed in the same way that a menu or report would be?

Environment type

There are a wide range of types of development environment, and the best type for one group of users is not necessarily the best for another group.

Mouse driven and menu driven environments are easiest to use and therefore suitable for novices, whereas command driven are the quickest to use by experts.

Is the application development environment menu driven, command driven, or a combination of the two?

Is a mouse driven environment available?

Code generation

Some application generators generate applications in COBOL or another third generation language. Ideally an application can be tested interactively and then either compiled or generated. Further, some products produce a form of tokenised code which is more efficient than interpreted code but less so than compilation.

Is the application generated, compiled, tokenised or interpreted? Can more than one option be used?

Some systems generate code in a range of languages. A standard language is desirable.

What languages can be generated? Do these comply with international standards?

Integration of editing and compiling

Where a specification needs to be compiled, it should be possible to initiate the compilation from the development environment (for example from within a specification editor); the developer should not be required to master the intricacies of job control languages or compiler steering lines.

Is the compiler or generator invoked:

- *directly from the application specification maintenance subsystem?*

- *using job control?*

Some generators always produce the entire system in a single generation run. For large systems this can take a considerable time and is not suitable in a large or volatile system environment.

What is the smallest component of the application system that can be separately generated?

Chapter 2
Usability

- *entire systems*
- *single on-line transaction or batch program*
- *other (please specify)*

The unit of generation is not necessarily the same as the unit of compilation. Ideally it should be possible to recompile individual transactions (exchanges or batch programs).

What is the smallest component that can be separately compiled?

- *entire system*
- *single on-line transaction or batch application*
- *other*

It is desirable that the user has some control over when recompilation takes place.

If the application specification is modified, is the application recompiled:

- *automatically at the next program execution?*
- *when directed by the user?*
- *at the completion of the edit of the specification?*
- *other?*

Speed of compilation turnaround	It is desirable that any compilation be quick so that the application is rapidly available for testing. Ideally, this implies operating in a foreground (ie real-time) environment in order to obtain a rapid response.
	AGEs which allow the rapid development of the application, but then require systems programmer

45

action to link the application to the database and/or transaction processing monitor are not productive. These options may be selectable by the developer. What is used may be dependent on the machine resource available.

Are compilations run in a real-time environment?

Is any non real-time processing involved before an application can be tested in an application system environment?

Development help information

At any stage of development it should be possible to obtain on-line help information; this facility is especially useful after the system has generated an unfamiliar error message. Several products now provide this facility. Some allow on-line access to the manual in a help mode.

Are help screens available as part of a comprehensive help system?

Quality of generation

Where code in a third generation language is generated it is important that it should be of a high quality.

One of the quoted advantages of a source language generator approach is that the code is easily maintainable.

Can the generator be constrained to produce code that conforms to recognised programming language standards?

What documentation, if any, is generated?

The form in which programs are generated can have an effect on machine performance.

Are the generated programs reentrant, reusable or single thread only?

Chapter 2
Usability

Some systems generate a single, very large application load module element - this is undesirable. Ideally the system designer should be able to fragment the load module so that the fragmentation is not apparent to the end user.

What control does the application developer have over the maximum size of the resultant application load modules?

If a data dictionary is not being used then it is necessary to separate development and production (object) libraries. Ideally the user should be able to nominate where the object code will be saved.

If the application is compiled, where is the object code placed and what control does the user have over placement?

Many of the interpretive systems semicompile their code. The vendors frequently claim them to be 'compiled' systems.

Is the resultant object code the machine code for the target machine?

2.4 Application testing

AGE produced applications, like conventionally produced applications, need to be developed and tested in sections as the application specification may contain errors. To achieve high productivity, an AGE must include testing and debugging aids.

Does the AGE include aids for testing, debugging and error correction?

Prototyping

In order to prototype the application it is necessary to produce a dialogue that at a minimum appears to function.

47

Prototyping is a useful facility to demonstrate to a user what a system is likely to look like (at the exchange/dialogue level).

Can menu hierarchies be quickly and easily implemented?

Do facilities exist for linking and displaying screen sequences before writing application code?

'Screen painting' is useful in a prototyping mode. If the screen definition is within the application then it is probably not very easy to have common screens in several exchanges.

How are the application's screens defined?

- *interactively, using a 'paint the screen' technique*

- *by drawing the layout using a text editor*

- *by specifying field names and coordinates on preformatted help screens*

- *by specifying field names and coordinates within the application definition*

- *by specifying field names and coordinates in the screen definition language of the host Transaction Processing system*

If screens cannot be displayed immediately (say they require compilation) then the AGE is less suitable for working directly with a user.

Does the screen definition require compilation before it can be displayed?

If the screen is defined using field coordinates, can the screen be displayed immediately?

Systems which use the host system's run time screen mapping facilities (and therefore invariably generate

Chapter 2
Usability

screen definition source language statements) are usually less suited to a rapid development/prototyping environment.

Does the screen definition process include generating source statements in the host system's screen definition language, followed by having to compile this definition?

Some systems will only pick up a new screen or program when they are restarted. For others, the program and screen format libraries are dedicated and cannot be updated. Such restrictions make rapid development, in a prototyping environment, difficult.

Does the system normally have to be brought down or restarted

- *to include revised screen formats?*
- *to include new screen formats?*
- *to include revised applications?*
- *to include new applications?*

Incremental testing

To allow incremental testing of an application, it is desirable for the AGE to provide sensible defaults for, as yet, unspecified attributes. If an unspecified module is invoked, the system enters 'test' mode, allows the tester to emulate the results of the module by adjusting variable values, and continues. Another approach is to allow program statements or segments to be executed immediately in isolation, but this is less convenient than the 'defaults' approach.

It is highly desirable that parts of an application can be tested in isolation, usually before other parts are developed.

Can applications be developed incrementally using a top down design philosophy (ie allowing testing with an incomplete application specification)?

Can application elements be invoked independently?

Can individual application specification statements be executed on entry? If so, can such a sequence of executed statements be saved to form part of an application specification?

Application area testing	While applications are usually developed on-line, it may on occasions be desirable to have the option of developing them (ie specifying and testing them) in a batch mode. Several products allow the specification, but not the testing, of applications in this manner.

Can the generator operate as a batch or pseudo-batch job?

Further, it may be necessary to simulate the functioning of a distributed system for the purposes of testing, to see if those elements of the application which specifically relate to that, are functioning correctly.

Is there a facility to simulate a distributed environment in order to test those areas of an application?

Debugging	In order to locate errors, it is helpful if the AGE contains some facilities to enable the run to be monitored or logged. Most products are weak in this area, requiring that the application specification be modified to obtain debugging trace information.

What facilities (please state whether interactive or not) exist for runtime debugging of

- *on-line applications?*
- *distributed applications?*
- *batch applications?*

In particular, a good debugger should include the following facilities.

Can the debugger display variable values?

Does it allow the setting of break points?

Does it provide step-by-step tracing?

Is it possible to view the source code during execution?

Systems could attempt to test for completeness by ensuring that all process inputs and outputs are compatible. Most systems do not really address this area.

What facilities exist for checking the consistency and completeness of the system specification? Which of these facilities are interactive?

2.5 Data and application definition

Most application generators make significant use of globally maintained data and file definitions; this avoids the need to respecify. However, restrictions on the use of these definitions will affect the product's usability.

Flexibility of global definitions

While centrally stored global definitions are useful, it is frequently necessary, at a local level, to override a global definition (for example for specific field validation or edit rules).

For some applications different validation rules, or slightly different screen layouts, may be required in different parts of the application.

Can definitions be stored centrally for global use? If so, can these global definitions be overwritten locally in order to provide local variants?

The same user, at different times, or different users, may require a different view of the data at the same point in the application. For example, the layout of a screen may change, or certain fields and their prompts may be suppressed for security reasons.

Can one definition, say of a screen, be stored with multiple variants?

Element specification procedure

It should be possible to specify elements (ie files, fields, processes etc) without regard to the order of specification (although it is accepted that some entry time checking will not be available if other elements are not defined). Products where the elements need to be specified in a strict order are not helpful; they necessitate analysts and designers working with paper when the dictionary should be a better, more productive vehicle. In addition, with such restrictions, changes to definitions become more time consuming.

Do the files, fields, exchanges and processes need to be defined in a set order, or can the developer impose his own construction sequence?

3 Functionality

The functionality of the product, in respect of developing applications, is a major factor in determining its potential productivity level. This applies both to the development process and the resultant application. Tools that can build powerful constructs enable the application to be built quickly and enable the end user to access data efficiently.

3.1 Specification storage

The mechanism by which the application specification is stored will affect the functionality of the tools which use it.

Application specifications stored in simple file or database structures can usually be modified with a variety of editors. For larger, more complex applications some form of dictionary is desirable. Application generators' own dictionaries are frequently unsuitable for general dictionary purposes unless they have been designed as comprehensive complementary products.

How is the application specification stored?

- *host operating system file*
- *conventional database file*
- *host operating system library file*
- *application generator's own file system*
- *application generator's own data dictionary system*
- *proprietary data dictionary system (please specify)*
- *other (please specify).*

3.2 Specifying language issues

AGEs vary widely in the ways in which applications are specified. Extremes vary from a pure declarative approach, through 'form filling' type languages, to more conventional looking programming languages.

Language type

Chapter 2 included a discussion on procedural and event-driven languages. The consequences go beyond those discussed there.

By definition, non-procedural languages do not have conditional constructs. However, as has been noted, most AGEs go only part way to providing non-procedural capability. Of these conditional constructs, that may be supported, the most important are:

Does the language support IF-THEN-ELSE, or equivalent?

Does the language support DO-WHILE loops, or equivalent?

Does the language support FOR loops, or equivalent?

Does the language support a CASE statement, or equivalent?

Another feature of some AGE languages is the ability to support recursion. This is the ability for a routine to call itself, repeatedly. This can be useful in certain types of application, particularly in handling tree structures. The drawback is that the program logic is likely to be complex and unsuitable for inexperienced programmers. On the other hand achieving this sort of programming without recursion, can also be complex.

Does the AGE support recursion?

Another factor with regard to recursion is the commensurate performance overhead. When a recursive program calls itself repeatedly, memory may need to be allocated on a dynamic basis for variables required within the recursive routine.

Chapter 3
Functionality

Does the AGE, in runtime mode, provide dynamic allocation of memory for use by recursive routines?

Is there any limit on the amount of memory that may be allocated in this way?

There are a number of other factors which should be included within the AGE if it is to be most efficient in development terms.

Which of the following data types are supported:

- *Boolean?*
- *Integer?*
- *Floating point?*
- *Fixed decimal/money?*
- *Fixed length character?*
- *Variable length character?*
- *Date?*
- *Time?*
- *Long text?*
- *Binary?*
- *User defined data types?*

What date and time functions are supported:

- *Add/subtract days?*
- *Add/subtract months?*
- *First of month?*
- *First of year?*

- *Extract day/month/year from date?*
- *Extract day of week from date?*
- *Extract month name from date?*

Does the AGE provide support for null values (that is, the field is empty as opposed to zero)?

How many types of null value does the AGE support?

What string operations are supported:

- *Concatenation?*
- *Substrings?*
- *String length?*
- *Convert upper/lower case?*
- *Trim trailing blanks?*

What data type conversions are supported:

- *String to number?*
- *Number to string?*
- *Date to string?*
- *String to date?*
- *Others?*

Object orientation — A distinction must be made between AGEs that provide object oriented features with regard to form design and other on-screen features, as opposed to those that are object oriented throughout. This is particularly true with regard to inheritance. These issues were discussed in detail in chapter 2.

Chapter 3
Functionality

At the moment, almost all AGEs that implement any form of object orientation, do it only for screen-based objects and not in the language per se. There are, however, a few exceptions. Selection of one of these may involve a philosophical or strategic decision that is beyond the scope of this volume.

Suitability for programmers

The type of specification language used, affects the experience profile requirements of the target application developers, as well as the likely acceptability of the product, especially by professional programmers. Some products contain constructs which are likely to be familiar to professional programmers, while others are usable by those with less experience (frequently at the expense of generality or completeness). In general, menu driven, form filling or application specification languages of very simple appearance, are more likely to encounter programmer resistance on the grounds that the building of applications is not 'programming'.

Some products generate parts of programs, rather than complete application systems. These may be termed program generators rather than application generators. Note, however, that most generators allow the inclusion of conventional code for exceptional processing. Some specification languages lack the required functionality to perform all the processing required.

Can the entire application be written in the specification language, or is it usual or necessary to use a conventional programming language for parts of the application?

Whilst a good application generator specification language is capable of specifying most commercial applications completely, there is always the possibility that some special process will require conventional programming techniques (eg unpacking data formats not supported by the generator). Commercial generators should support standard languages.

Is it possible to specify the invocation of subroutines or procedures written in a conventional programming language from within an application? If so, which languages are supported?

Style and quality	Good, well designed languages have few irregularities or exceptional rules for the programmer to contend with. Some products are close enough to conventional languages to be easily mastered by programmers, others have specification languages based on different concepts such as set based constructs, or object orientation, which some conventional programmers may find hard to adjust to. Non programmer languages are usually more rigid in what they allow the developer to write. A variety of approaches have been adopted for specifying applications. Short annotated examples of actual applications should be sought. Examples taken from training manuals will usually be satisfactory.

Please describe the style and quality of the application specification language, using examples if this is beneficial?

Data sub-language	Application Generation Environments are most often used with Database Management Systems. The AGE will use a data sub-language (for data definition and data manipulation) to interact with the DBMS. This is described in detail in Chapter 4.2.

Validation and integrity	The AGE does not necessarily need to access the DBMS for validation and integrity purposes. The two principle concerns are referential integrity and cascade deletes. The former provides a check to ensure that data linkages are maintained under all circumstances. The latter is similar in that it ensures the preservation of integrity in data structures. While it is possible to program around these, it is a lengthy and complex task. It should be noted that the best method of implementing referential integrity, is through the dictionary, where it is defined just once.

Is referential integrity enforced through the dictionary?

Is referential integrity enforced through database triggers?

Is referential integrity enforced through forms?

Are cascade deletes supported?

3.3 Components

There are a number of elements of an AGE which do not comprise either the language, or immediately associated tools such as editors. While this volume does not seek to review all of these in detail, their functionality may be pertinent to the evaluation of an AGE, especially where they form part of a bundled product set.

Large object management

The management of binary large objects (BLOBs) is likely to be under the control of the language and DBMS, and does not strictly fall under the heading of a component. However, it is likely that BLOBs will relate to some special-purpose requirement, and they are therefore treated here as if they were an independent entity.

Some products, both AGE and DBMS, can handle long text but not BLOBs. The former is strictly speaking a subset of the latter. Precisely what data types are supported was covered (above) in the section on Language type. The only remaining issue is the extent of manipulation provided.

Of those products that can utilise long text and/or BLOBs, few can do anything other than store them. This means that in order to access a particular part of the BLOB, the whole thing has to be read. This can be very resource inefficient since BLOBs of, for example, video, can be many megabytes in length. A method is needed to break up the BLOB. Only a handful of products are able to do this.

Does the AGE, or its associated database, provide a mechanism for indexing BLOBs?

If so, how is this achieved?

Menu management

An important issue with regard to building an end-user menu system is access control, which restricts users to seeing only what they are allowed access to. Similar facilities may also be provided with regard to terminal and time-of-day. These, collectively, are sometimes referred to as dynamic menus.

Does the AGE provide user-sensitive menuing?

Does the AGE provide terminal-sensitive menuing?

Does the AGE provide time-sensitive menuing?

Forms management

AGEs are increasingly moving towards a forms-oriented development environment. There are a number of features that should be included in a good product of this type, most of which are self-explanatory. Many of these relate to environments utilising GUIs.

Does the AGE provide support for scrolled areas?

Does the AGE support horizontal scrolling?

Does the AGE support updating of fields within a scrolled area?

Is there a field zoom capability?

Is windowing supported?

Chapter 3
Functionality

Can window data be updated in situ?

Are cut and paste facilities provided?

Is there a screen painter?

Does the screen painter provide windowed access to the dictionary?

Can graphics or images be included in forms?

Does the language allow functions to be attached to function keys?

What level of text handling and/or word processing is provided by the AGE?

One of the most common requirements in developing forms-based systems is that of providing logical links between different fields.

Is there support for cross-field validation?

Event driven AGE languages provide points at which specific procedures may be inserted once a particular event has occurred. Pre and post-map procedures run before and after a mapping process, typically at the field level.

Does the product support both pre-map and post-map code entry points?

Capabilities of query facility	The increased requirement from end users to be able to access data directly, means that some form of powerful query facility is frequently required to augment the purpose built application constructed with an AGE. Note that many query languages are particularly suitable for use by unskilled end users.

61

Major DBMS products are associated with Report Writers and Query Languages. AGEs built on such products are likely to meet most requirements in both areas.

Are queries expressed

- *using a formatted help screen?*

- *using a relational or set based language?*

- *other (please specify)?*

Does the Application Generator include a query language facility?

Does the query language adhere to the SQL international standard? If not does it adhere to any other generally accepted standards? Please specify which standard.

Can individual query language statements be executed interactively?

Some systems differentiate between their own file systems (which are supported) and alien file types that are not.

Does the query language facility operate with all the file or database structures supported by the application generator? If not, which are not supported?

A join type facility within queries is necessary for all but the most simple application systems.

Are queries restricted to single files only?

General information - note that systems using a formatted help type screen will possibly have a second interface for direct use within applications.

Does the user have to know the files in which the data items reside or is the information held within a dictionary/directory?

Can the records or fields retrieved as a result of executing a query, be saved and available to be used in further queries?

What facilities are available to assist the end user to formulate queries?

What facilities are available to make the breaking down of complex queries easier?

Can retrieved data be formatted for presentation using the report writer facilities?

Can graphs, histograms, pie charts and other graphical representations be produced?

Capabilities of report writer	Report writing is a significant part of most batch or offline applications. Real applications often demand complex reports. Good report writers, by enabling that function to be separated, simplify the system and give useful productivity gains.

General information - note that field/coordinate type specification is the least usable. The use of a separate report writing package may imply that the application generator produces a separate, intermediate database/file.

Does the application generation environment include a report specification facility?

How are report formats specified? Note that some products may offer more than one option.

- *interactively using 'screen painting' techniques*

- *interactively by means of graphical point-and-pick features*

- *by specifying field names and coordinates on pre formatted help screens*

- *by specifying field names and coordinates within the application*

- *by specifying field names and coordinates to a separate report writing package*

- *other (please specify)*

'Screen painting' is good when working directly with the user, but usually the definition requires editing later for consistency. It is undesirable to have to repaint the screen to effect minor changes. When working directly with a user it is desirable to be able to show them the report layout quickly.

If a 'screen painting' technique is used, can the definition be edited?

If the report is defined using field coordinates, can the report layout be displayed immediately?

Regardless of the means of defining the report, does this then generate code which can be edited?

General questions on report writer's capability.

Does the report writer allow access to the dictionary during report creation? How is this achieved? Does the report writer make automatic use of data relationships defined within the dictionary?

Does the report exist as a separate entity independent from the application specification?

Can the report specification be used within multiple applications?

Can report output be directed to the VDU initially during development and then to a printer for production operation?

Chapter 3
Functionality

What options exist for sorting the data to be reported?

Can ascending or descending key sequences be specified?

What is the maximum number of sort keys that can be specified?

Can ascending and descending sequences be intermixed?

Can control breaks be specified? If so, what is the maximum number?

What facilities exist for totalling and other computation?

Is it necessary to specify field formats or column headings within the report specification? If not, from where are the defaults obtained?

Can default field formats or column headings be overridden in specific instances?

Can graphs, histograms, pie charts and other graphical representations be produced?

Can labels be produced?

It may be desirable to produce more than one report from one pass of a file, for efficiency and integrity (of results) reasons.

What is the maximum number of reports that can be produced from one pass of the data?

3.4 Environment type

AGEs are falling more and more into two camps. Either they are GUI based or they are not. In the latter case they will be either menu driven or command driven. Increasingly, products are moving into the first camp. Most of the issues involved with using either variety have already been discussed, except for developing applications which themselves will run in a GUI environment.

Graphical user interface

If a developed application is to run in a GUI environment, the first question to be asked is which commercial, and preferably standardised, GUIs are supported.

Which of the following GUIs does the AGE run under:

- *Windows?*
- *Windows NT?*
- *OS/2 Presentation Manager?*
- *Open Look?*
- *OSF/Motif?*
- *X/Windows?*
- *Others (please state)?*

The other consideration in using a graphically based AGE, is which of the tools used in the graphical environment can be easily incorporated into the application under development. This will most particularly apply to forms.

Which of the following can be incorporated within a form:

- *horizontal and vertical scroll bars,*
- *status line,*
- *title bar,*
- *menu bar,*
- *help status,*
- *control-menu box,*
- *command buttons,*
- *radio buttons,*

Chapter 3
Functionality

- *highlights,*
- *check boxes and so forth?*

How is this achieved?

What facilities are provided for iconisation of applications and functions?

What facilities are provided for incorporating pull-up or pop-down menus, and for accepting mouse-based user operations?

Appraisal and Evaluation Library
Application Generation Environments Volume

4 Other system components

When assessing software for a project, it is not reasonable to assess the capabilities of an application generation environment in isolation from the capabilities of other components of system software. This is true irrespective of whether the components are supplied as separate products, or as part of the AGE. Other major system components, aside from those already discussed in Chapter 3, are:

- DBMS
- TP (Transaction Processing) Monitor
- Data dictionary
- CASE tool

It may often be necessary for an AGE to interface to other data management products that are possibly being installed as part of the same project. These products may alternatively have been installed as part of a departmental strategy or standard, or the AGE may be required to extend an existing application that uses other products.

The ability to interact with other products reduces the possibility of having to install software that duplicates existing capability. It is likely to be cheaper in hardware and software terms than separate environments.

AGEs are marketed as:

- single packaged products with integrated data management capability
- part of an 'Integrated Product Range' from a single supplier, with the individual components within the range sold as separate products

- having interfaces to host system data management software.

Where products need to interact with other components, the interaction is particularly important with respect to those system components mentioned above, although integration with the operating system will also need to be considered.

Note that if separate appraisals of other components are being undertaken then the integration aspects may need to be considered in isolation because of the multiplicity of combinations that may be possible.

4.1 Capabilities of DBMS

The facilities of a DBMS include a mandatory requirement to support multiple updating users in an on-line environment. Some products are not supported by proper DBMS software despite the claims of the vendors. Suspect areas are usually insufficient support of concurrency control, or recovery in a multi user environment.

Is the AGE supported by a fully functioning DBMS? How does the DBMS support concurrency control and recovery in a multi user environment?

4.2 Integration with DBMS

If the requirement is to use a host DBMS, the code produced by the AGE must be capable of utilising and exploiting the facilities provided by that DBMS, in order to support recovery and integrity and to provide usability and functionality. Many products do not have a particularly close interface with alien database management systems. This may be significant for those projects considering using open systems interfaces. Interworking may be possible but fall short of well integrated, easy to use facilities. For example, it may be necessary with some products to write interface code in a conventional language to perform functions like opening and closing a database and to

Chapter 4
Other system components

handle the mapping between database records and the record structure used by the application generator.

Standards conformance

Some vendors claim that their AGE and DBMS products support ANSI and ISO international standards. Support of such standards should greatly aid integration between AGEs and DBMSs from disparate vendors.

By conforming to appropriate standards will the AGE be able to exploit the facilities offered by a host DBMS from a different source claiming to support the same standard?

What standards does the AGE conform to?

Structured query language

The use of a structured query language (SQL) is increasingly the main means of querying a database. Although SQL originated as a relational database front-end, it has proved so popular that many mainframe databases, which were primarily designed on hierarchical or networked models, have added SQL to their functionality.

The extent to which SQL is supported varies widely between AGEs and there are several pertinent questions to ask.

Is SQL supported?

If so, what standard of SQL - ANSI? ANSI level 2? IBM's SAA?

Does the AGE language use SQL explicitly?

Does the AGE language generate SQL?

Can SQL be generated from point-and-click methods in a GUI?

Some products provide support for interactive SQL. This allows direct access to the database and can be especially useful during system testing, providing a quick way of querying data.

Does the AGE support interactive SQL?

Database and file definition

Database and file definition is usually carried out by using the DBMS tools, or by using a Data Definition language. Definition from within the AGE can be beneficial.

The application must clearly know of the database/record structure. Ideally this should be held in a dictionary/directory. Some systems describe the data structure as part of the application. This means that if the structure changes all applications using the structure require amendment. Some systems deduce a database structure from the application screen layouts.

Is the database or file definition provided as part of the application specification

- *explicitly?*

- *implicitly?*

- *not part of the application specification?*

If the DBMS and the application generation environment are integrated, then they should use the same definition. If they are not integrated it is useful if the database definition need be specified/coded only once.

Is the database or file definition used by the application, specified in the same language as the database or file description used by the DBMS?

Chapter 4
Other system components

If the application specification implicitly produces the database definition it is probable that the user has little or no capability to tune the structure.

Can the same file or database specification be used by more than one application? Is the database definition separate from the application specification?

If the file or database definition is derived from, or contained within, the application specification, what facilities exist:

- *for tuning the database structure?*

- *for reorganising the database structure?*

Gateways

The AGE may be part of an integrated product range, it may use only standard system files, or it may be sufficiently flexible to access a range of DBMS products. This requirement will be dependent on the project environment.

Some systems are primarily designed to interface to their own DB system. They do, however, handle other database structures, but in read only mode. Frequently when an application generator is able to process an alien database structure it does not support all of the DBMS's capabilities.

Is the application generator primarily associated with the vendor's own database system?

What other files and database systems may be accessed?

What restrictions are imposed when accessing data from other systems?

- *read only*

- *limits on the structured complexity of the database*

- *database or file features or operations not supported.*

Application generation environments sometimes support only a small number of data representations. Conversion is usually by a conventionally coded interface routine where standard facilities are not already provided (see chapter 3). If the generator is to be introduced into an existing environment the inability to handle any general database structures produced for conventionally coded (existing) applications is a severe limitation.

Can the generated application access files and databases previously created for conventionally programmed applications? If so, how are data types converted when they are not supported by the application generator program?

A 'CALL' type interface is necessary if conventional applications are to access the data held in the DB component of the application generation environment. If such an interface does not exist all applications needing to exploit the DB environment will have to be implemented using the AGE. Thus the user will be 'locked in' to the AGE system which could cause problems if runtime efficiency is an issue.

If the application generation environment is primarily associated with its own database system, how can the data be processed realistically by conventionally produced applications?

Validation and integrity	Where the AGE is integrated with its own database, referential integrity may be left to the DBMS. It will be a limiting factor if this is the only way this may be achieved. This issue was addressed more fully in chapter 3.
4.3 Operating system	The operating system under which the AGE will run will, necessarily, limit the choice of available product. It will therefore be necessary to establish that potential products run on the chosen platforms and operating

Chapter 4
Other system components

systems. If this is to be an open system then Posix compliance is a requirement:

Is the AGE Posix compliant?

If the AGE is not Posix compliant, which operating systems will the AGE run under?

Recovery

Recovery for multi exchange business transactions, is required if general systems are to be constructed. Ideally, it should use a DB provided delayed update technique, and specifically a two-phase commit for distributed processes. Some systems provide their own facilities to enable the user to directly manipulate temporary data storage areas. Minicomputer based systems may require the user to create a small temporary file for the life of the transaction. Note that any such temporary data storage area should be recoverable.

Does the AGE rely on the underlying DBMS for recovery?

If the AGE does not rely on the underlying DBMS for recovery does it support recovery for multi exchange business transactions? If so, are database updates performed using:

- *DBMS handled delayed update technique? Two-phase commit, or otherwise?*

- *AGE handled delayed update technique (using the temporary data storage area)?*

- *application handled delayed update technique (the application specification saving data within the temporary data storage area or other files)?*

- *locked database records across exchanges?*

Does the application generation environment contain its own data file handling routines? What information is logged automatically for recovery purposes:

75

- *before images?*

- *after images?*

- *database transactions?*

- *other (please specify)?*

What utilities exist to facilitate database recovery?

For batch applications, is checkpointing or any other recovery technique supported?

If only simple use of DB/TP recovery services is made, recovery probably will be to the last (completed) exchange. In many systems (especially secure ones) this will not be sufficient - the operator may have left the terminal when it becomes live again. Ideally the application should be able to decide the recovery state.

In the event of a system failure during an on-line or distributed application, will the user's terminal be recovered to:

- *the end of the last (completed) exchange?*

- *the start of the dialogue?*

- *the logon sequence only?*

- *a point determined by the application design? (for example start of business transaction).*

Distributed capabilities Given the current impetus towards client/server methodology, all vendors are likely to suggest that their product supports it. There are, however, a number of issues to be clear about.

Which operating systems are supported for servers?

Which hardware platforms are supported for so-called open operating systems?

Chapter 4
Other system components

Is Posix supported?

Which operating systems are supported for clients?

Can dumb terminals be used to develop applications in the same manner as PCs or workstations?

Can multiple operating systems, at both the client and server level, be mixed on the same system?

If so, are there any limits?

Is there any impact on recovery, for multi-system client/server operation?

Is there any impact on access security, for multi-system client/server operation?

Other factors, such as two-phase commit, and whether the AGE supports concurrent access to more than one type of database, have already been addressed.

4.4 Capabilities of the TP monitor

In some environments a TP monitor will be required to control terminals, both for service and recovery purposes. Special support is required to handle multi phase (multi exchange) business transactions, as communication areas may need to be preserved, and database records locked, over several exchanges. Very few products are able to do this.

In cases where there is a choice, a product that uses a TP monitor is likely to give better performance than one that does not.

Can the AGE be used with a TP monitor? Is the TP monitor, or the AGE, restricted in any way when used together?

4.5 Integration with transaction processing system

In some environments the development of large or complex computer systems is likely to necessitate the use of advanced transaction processing monitor facilities for performance reasons. This section only applies to this type of system.

Few AGEs have these facilities built in, so it is often required to use some host TP monitor. In order to make the best use of its facilities, the AGE needs to be well integrated with the TP monitor. In addition, the degree of integration at the interface level affects the usability of the AGE.

Standards conformance

Historically, TP Monitors have tended to be proprietary products running exclusively on mainframe systems. With the advent of increasingly powerful Unix processors it is now realistic to consider running OLTP (online transaction processing) on such a platform. While there are no particular standards that currently apply to TP monitors themselves, there are some vendors of TP-specific AGEs, who require any accompanying database to be XA compliant (see Chapter 1). This means that the DBMS complies with the X/Open two-phase commit protocol. Note that TP monitors do not necessarily run in conjunction with a DBMS.

Does the AGE only support XA compliant DBMS products in conjunction with its supported TP Monitor(s)?

AGE integration

As has been noted, some AGE suppliers are now offering specific front-ends to support certain TP Monitors.

Does the AGE have a specific front-end to support OLTP?

What additional facilities does it offer?

Which TP Monitors are supported?

Chapter 4
Other system components

Application integration	For high performance applications it is essential that the application be well integrated into a TP environment.
	Can the application run in a TP environment? Does it make full use of the TP monitor facilities?
	How well is the application integrated into the TP environment?
Development environment integration	To develop an application to run under a TP monitor it is possible that the development environment will also need to utilise one.
	Can the development environment run under a TP monitor? Is it necessary to produce a TP application?
	To what extent is the development environment well integrated into the TP environment?
Recovery	Recovery is usually managed by the DBMS rather than the AGE. When this is not so, coordinated DB/TP recovery is necessary. It is probably acceptable if the generator uses a host DB/TP system. This may not be the case if the generator contains its own DB system.
	For on-line systems, is database and TP recovery coordinated?
	Does the application generator contain its own terminal/message handler? If so, what information is logged automatically for recovery purposes?

- all system accesses
- all input messages
- all successful transactions
- all output messages
- other (please specify).

Does utility software exist to enable messages to be reprocessed?

4.6   Data dictionary   The AGE data dictionary, whether provided as a part of the AGE, or interfaced to as part of a separate DBMS, is absolutely crucial to the success of the AGE. It is not unreasonable to say that it is probably the single most important part of the application generation environment. Since the dictionary specifically controls and defines that environment, then if the dictionary is impoverished so too is the AGE.

Data dictionary capabilities

Good data dictionary facilities are usually regarded as central to effective documentation and control; this applies to AGE and non AGE developed systems.

While many AGEs include some form of data dictionary capability, this is frequently unusable for non AGE applications.

Can the data dictionary be used for non AGE development?

It is important that the meta data (data about data) in the dictionary can be accessed in an ad hoc manner to produce additional reports required by the analyst. This can be best achieved if the data dictionary has been developed using tools available to the analyst to extend the dictionary if required ie it has been implemented using an AGE and DBMS.

Is the Data Dictionary implemented as a database that can be accessed by standard DBMS tools?

Does the dictionary tool set include a range of standard reports?

The dictionary should be capable of holding more than just the data definition. Additional detail is required for full support of analysis, and potentially for

Chapter 4
Other system components

automatic systems generation. Definition of referential integrity within the dictionary, has already been discussed

.Does the dictionary support process as well as data definition?

Can forms be stored in the dictionary?

Does the dictionary support the business model as well as the application model?

In particular, a major consideration will be the type of dictionary employed. These fall into three categories: passive, active and dynamic, though some products may offer hybrid versions.

A passive dictionary simply stores data on the objects used by the AGE language. An active dictionary adds the ability to hold linkage information between related objects, such as which records use a particular field definition. This allows the developer to perform impact analysis, to determine the consequences of any proposed changes. Finally, a dynamic dictionary goes a step further by incorporating a link with the run-time environment. If an object is modified, the system knows it, and with some products this means that programs may be recompiled. Dynamic dictionaries may degrade performance in the development environment, but this is not necessarily the case.

Is the dictionary passive, active, dynamic or a hybrid (state of which)?

If the dictionary is dynamic, does this have any performance overhead?

Most dictionaries provide the facility to assign different attributes to fields, for use in forms or record definitions. Such data validation should be definable within the AGE and stored in the data dictionary, since this will mean that they only need to be created once. The following are the most significant.

Can a field be mandatory?

Can a field be defined as numeric?

Can a field be defined as alphanumeric?

Can you define range checks on a field?

Can you define a field as belonging to a specified set of values?

Can you define pattern matching (ie fixed length, specific positioning for alpha and/or numeric characters)?

Can you incorporate formula checks (eg a check digit routine)?

Is there automatic look-up (ie specification of a file access to check a value)?

Can you cross-validate with other fields?

Is there support for default values?

What action occurs if an input field fails one of the validation checks?

Can default actions (as opposed to values) on input be defined?

What data output representations can be specified when defining a data item:

- *default column headings*

- *default size (different from storage size)?*

- *edited output representation (please indicate how the output representation is specified)?*

Chapter 4
Other system components

Within a distributed environment it may be necessary to have a distributed dictionary if development is split across more than one geographical location. This was discussed in chapter 1.

Integration with data dictionary

Application generators, almost without exception, use some form of dictionary or directory as a repository for system specification information. The role of a data dictionary and the need for an AGE to use centralised element definition, make it desirable that the AGE and other parts of the organisation use the same dictionary. The sophistication of the interface between the AGE and the dictionary affects the product's usability and the ability to audit and control system development.

Some products maintain their own specific dictionary giving the organisation the problem of maintaining consistency between two different, non compatible dictionary systems. Some vendors have software bridges to overcome this problem.

Other vendors have powerful dictionary products with close interfaces to the AGE products. They are capable of supporting both analysis and implementation details and allow the dictionary meta structure to be extended to cater for local requirements.

A different approach, adopted by some vendors, is to utilise the dictionary of the host database, even though this may be derived from a different supplier. This is becoming particularly prevalent in the client/server environment though there are still some vendors who provide bridging techniques between distinct dictionaries.

User sites will frequently have a considerable investment in a proprietary data dictionary system and will wish to use the same system to volume (at least in parts) generated applications.

Does the AGE provide its own data dictionary?

Does your proprietary DBMS use the same dictionary?

What other vendor (DBMS) dictionaries does your AGE support?

Does it do this by means of a bridge between the dictionaries or does it directly use the other vendor (DBMS) dictionary?

If the latter, does it simply use the (DBMS) dictionary or can it provide extended facilities?

If the latter, what are these extensions, and how is this achieved?

4.7 CASE tools

CASE tools fall into various categories, depending upon the tasks they are required to fulfil. Whatever these may be, they are increasingly important in the development of applications, and should complement the selected AGE.

It should be noted that the view taken in this volume is that addition of a CASE tool to AGE, is what makes into an Application Development Environment. It is obvious that the distinction between CASE and AGE products is becoming increasingly blurred in some areas. The approach taken is that CASE tools are primarily concerned with the systems analysis and planning of a project, whereas the AGE is about actual generation of an application. Further guidance on CASE tools can be found in the *CASE Tools* volume of the *Appraisal and Evaluation Library*. Users of SSADM should consult CCTA's *SSADM Tools Conformance Scheme* volumes.

CASE tool capabilities

This is not the place to examine CASE tools in detail, but the following questions will broadly define the scope of any such product.

Is it a lower, upper, or meta CASE tool?

Does it generate code?

Chapter 4
Other system components

Does it incorporate reverse engineering?

Is there a metrics facility available?

Method support — Any CASE tool selected should support the method used by the organisation. There is, however, an issue to be considered with regard to object oriented methodology, if an AGE is selected which uses an object oriented approach throughout. This will be discussed further in Chapter 6.

CASE integration — The CASE tool should use the same dictionary as the AGE. In practice this will considerably limit the choice of CASE tool (assuming the AGE is selected first).

Which CASE tools are available which can share the dictionary used by the AGE?

Which of these support our method?

Often vendors of AGEs may recommend particular CASE tools with which they have experience. It may be useful to request such information.

Ideally the dictionary used for a CASE tool needs to be extensible. That is, to allow users to define objects which can be stored in the dictionary, which are used as a part of the application building process. Such objects include typical CASE constructs such as entity diagrams, data flow diagrams and so forth.

Is the dictionary used for both the CASE and AGE products, extensible?

Appraisal and Evaluation Library
Application Generation Environments Volume

5 Efficiency

One of the primary objectives in purchasing an AGE is to reduce the time taken to develop applications.

In achieving this, AGEs necessarily consume a high level of machine resources, both during development and in production use. The performance/efficiency factors are usually of major importance.

It is also important to be aware that the measurement of the efficiency of an AGE may not be simply a matter of measuring CPU resources used. Other metrics include:

- Memory usage - some systems are not effectively multithreaded, and so require significant storage resources

- I/O activity - systems claiming on-line interactions with a dictionary usually generate much I/O activity.

- Network activity - distributed systems have a network component that may be significant. In addition, the location at which distributed functions take place, can also be a significant factor.

When an organisation acquires an AGE, it frequently acquires other advanced software such as a DBMS. Such software is usually resource intensive in its own right.

5.1 Development productivity

Vendors claim significant productivity gains for their products over conventional 3GL development. It must be stressed that these gains only pertain to a small element of the system life cycle, namely application design, and that use of an AGE does not remove the need for thorough systems analysis. Secondly a large proportion of any gain can be lost by iterating the design, particularly if a prototyping approach to application developement is used.

It is essential that any gains can be verified by contacting users of the product.

What productivity gains have been <u>objectively</u> measured by reference sites over the use of a 3GL such as COBOL or PL/1?

5.2	**Development resource usage**	Most AGE development work is carried out on-line and response time, and therefore machine efficiency, are important in order to maintain the productivity of application developers. While application development is principally a one-off activity compared with the application being run in a production mode, it is desirable that not too many resources are used. In particular, compilers for AGEs can be large. Also, because the normal method of working is to develop small sections of code, to test and debug them, and then to proceed, it is desirable that the application can be compiled/interpreted in small units.

The development resource requirement is often greater than for runtime of the developed end application. It may be that the project will have to accept slower development, or shift working, to fit hardware constraints. The requirement is also often greater than the vendor would care to admit. Again reliable reference site contacts are vital.

What hardware configuration is required (in terms of RAM and Disc Memory etc) to support the proposed number of application developers? Can this be verified by reference sites?

5.3	**Runtime resource usage**	Efficiency of machine resource usage for the developed application affects the type of application for which the product will be suitable. Interpretive implementations are usually less efficient than compiled

Chapter 5
Efficiency

implementations. This is no disadvantage in low transaction rate, test or prototyping environments. It is a serious disadvantage if high transaction rates are to be supported. Few products claim to be able to execute the same specification either interpretatively or from compiled code. A desirable additional requirement is the ability to invoke low level code, on occasions, for particularly performance critical sections of applications.

Separating the AGE and DBMS processing loads into a client/server architecture can give considerable performance gains. This will relieve the DBMS or server machine of the application load thus releasing resources for the DBMS.

Can the application front end be run on a separate processor from its DBMS? Can this front end be a workstation for each user?

Can more than one application front end be supported?

Is there an algorithm to estimate the resource requirements for each user? Is there a minimum memory requirement for each user?

Runtime usage is heavily dependent on the performance of the underlying DBMS and its query optimiser. However, not all DBMS have very efficient optimisers. At the least the optimiser should use gathered statistics and be syntax independent. This is even more critical in a distributed environment. If a query comprises making a join across ten records in one database and a million in another, it is essential that the activity takes place in the latter network node. It is also important, where heavy resource usage is required by a process, that the developer or end-user, as appropriate, be informed that this exceeds a specified threshold, and does he/she really want to proceed. It may even be necessary to prohibit certain activities by means of threshold levels and user permission levels.

89

What sort of query optimiser is provided by the DBMS?

Is there a distributed optimiser in the DBMS?

If so, what is its functionality?

Can developers be notified of resource requirements for a particular task, before execution?

Can end-users be similarly notified?

Is it possible for the DP manager to set threshold warning limits?

Can these be used to prevent unauthorised users/developers from undertaking actions which exceed the relevant levels?

Metrics

Metrics, as was mentioned above, apply to all the environments discussed above. In particular, there are metrics tools available, usually in conjunction with CASE products, that enable estimates of resource and productivity requirements prior to a project actually starting.

Is there a metrics tool available with the AGE or associated CASE tool?

If so, describe the facilities it offers for estimating both development and runtime productivity and resource utilisation.

6 SSADM

There are a multiplicity of design methodologies which, although principally concerned with CASE tools, also reflects on any selected AGE.

The use of SSADM is widespread in both private and public sectors. These users do not wish to introduce a different range of techniques and procedures for each Application Generation Environment they use, because that would compromise their standards, reduce protection of investment and generally increase long term costs.

With this in mind CCTA has recommended to government departments that they request suppliers to provide guidelines on how best to use their products with SSADM. This guidance should address all aspects of the product, including documentation and training The detailed objective of this CCTA policy is to:

- reinforce the fact that analysis and logical design still need to be conducted rigorously and in a controlled way, even when AGEs are used for implementation

- allow a smooth transition from logical to physical design without gaps, duplication or damaging compromise of requirements

- make full use of SSADM requirements and design deliverables - particularly in a turnkey or facilities management situation

- produce good quality documentation for the maintenance, future enhancement and eventual replacement of the system.

SSADM is presented and documented as an integrated set of structural, technical and documentation standards. The assessment of supplier support for their product in a project using SSADM should be assessed against all three sets of standards.

Compatibility

The following section explores the extent to which the AGE suppliers guidance, is compatible with SSADM.

The purpose of the SSADM Analysis phase is to analyse and document the user requirement without the constraints imposed by any particular implementation choices.

Does the AGE supplier provide guidelines on how to implement the application using the SSADM products?

Is there a tailored form of the SSADM Structural Model to lead the practitioner through the method?

The guidance should describe the use of standard SSADM end products and techniques with the modified techniques and end products designed to maximise the usefulness of the AGE. The guidance will be of little value if it cannot be quality assured with the standard SSADM product set.

Are there supporting descriptions which clearly cross reference the modified end products to standard SSADM end products, to assist the smooth and effective operation of SSADM Reviews?

SSADM is an evolving standard and suppliers of AGEs should be prepared to commit themselves to supporting new versions as and when they become available.

Does the AGE supplier intend to support future versions of SSADM?

Tool support for SSADM

If an analyst workbench is used, then it is likely that the results of the SSADM analysis will be stored within

Chapter 6
SSADM

a data dictionary, analyst workbench or CASE tool. The vendor may then supply tools that take analysis information from the dictionary or tool and automatically convert this into application outlines or even program code. The availability and sophistication of such tools is likely to have a significant impact on productivity.

In the assessment of such tools, the following points should be considered:

- the use of an integrated data dictionary to record the results of SSADM analysis activities, and cross check them with AGE design activities

- the use of an integrated, close or loose coupled analyst workbench to produce SSADM end products

- the ability of the tools to produce SSADM requirements documentation for the maintenance phase of the life cycle

- the SSADM Tools Conformance Scheme will test CASE tools' level of conformance to SSADM.

Are there tools available to map from analyst workbench or CASE tools into application code? Do such tools include links from the products of SSADM?

Have any such tools been assessed aginst the SSADM Tools Conformance Scheme?

CASE support for SSADM

The topic of CASE tool support for SSADM is not covered in depth within this volume. Further guidance on CASE tools can be found in the *CASE Tools* volume of the *Appraisal and Evaluation Library* and the *SSADM Tools Conformance Scheme* documentation.

Any CASE tool selected should support the SSADM methodology in its entirety. There is, however, always a delay between announcement of an updated version of the methodology, and its implementations.

Which version of SSADM is supported?

Which features does the tool support?

If not, will the missing features be added?

If SSADM support is not for the most up-to-date version, when is this expected to be released?

Other methodologies	It is conceivable that situations will arise where SSADM is not entirely suitable. In such circumstances it is desirable to use SSADM wherever it is feasible to do so, and the appropriate alternative only where strictly necessary. A suitable CASE tool should therefore not merely offer SSADM but also the ability to select elements from different methodologies and meld them together.
Object orientation	It is unlikely that any alternative methodology will offer any benefit over SSADM if used in the normal way. This may, however, no longer hold true if OO techniques are to be used and, in particular, if an Object Oriented AGE has been selected. In this case it may well be pertinent to mix an OO methodology (e.g. HOOD, Booch, Rumbaugh or Coad/Yourdon) with SSADM. As there is no widely agreed set of OO concepts beyond the basics, this raises the risk of becoming locked into a proprietary approach.

Does the CASE product support both OO and SSADM methodologies?

Is it possible to merge elements from both methodologies to provide a consistent approach?

Are there any limits imposed by the tool with respect to achieving this?

7 Quality and control

Rapid development of code and data structures requires that the AGE provides good control and documentation of the resources (files, databases etc) used. This usually requires that the AGE interfaces closely with the organisation's data dictionary and that the dictionary has sufficient access control and 'status' functionality.

7.1 Documentation

Documentation should be 'active' rather than 'passive'; ie the documentation should result from building a specification, not be a separate activity. Only in this way is consistency between documentation and application assured.

Can documentation of the application be generated as a by-product of building the application?

7.2 Development control

The application generator, or its supporting dictionary, should allow multiple versions of the application components to exist concurrently; typically development, production and historic versions. This facility exists in those products supported by a good data dictionary, but few others. Otherwise, projects need to maintain separate production and development libraries themselves.

Separate libraries are essential in a realistic environment. Individual projects, users and production systems must be protected from each other. Note that some systems have a single central library but that this is effectively partitioned. A single library is likely to grow very large.

Does the generator allow the definition of separate development and production libraries?

On-line interactive development can be expensive in terms of computing resources. Frequently, costing information relating to individual users/projects is required to provide a means of controlling access.

Does the application generator environment provide accounting information to enable the costing of application specification maintenance and development?

Version control and configuration management tools are increasingly in demand to control the development process.

Configuration management, in particular, requires more than just a simple check-in and check-out procedure. It will commonly be the case that a number of library items may need to be checked out simultaneously. Thus there is a requirement to be able to logically group development objects. Further, there must be procedures to ensure the consistency of objects checked back into the library. For further guidance on configuration management see the *IT Infrastruscture Support Tools* volume of the *Appraisal and Evaluation Library*.

Does the application generation environment provide mechanisms for managing 'versions' within 'releases' of the generated application?

Does the AGE provide configuration management facilities?

What facilities are included to provide logical groupings, and to ensure consistency?

7.3 Audit control

Because of the ease with which programs may be amended, it is essential that there is recognition of 'production status', where access to programs is strictly controlled. In addition, it may be desirable for audit and control purposes, that the person seeking access be identified and the date and time be recorded.

Chapter 7
Quality and control

It is likely to be an auditor's requirement that at least production libraries are protected against unauthorised access. Passwords are commonly used. Some systems rely upon facilities provided by the host operating system (for example separate files, each with the operating system's password access control mechanism applied).

How can the production libraries be protected from unauthorised use?

It may be an auditor's requirement to know when a program change has been effected. Ideally the system should write the date, time, user-id and program-id to a system log.

What information is logged when an application specification:

- *has the source accessed?*

- *has the source amended?*

- *is recompiled into an application program?*

Configuration management facilities are also applicable here, as they were with development control.

Are there configuration management facilities pertinent to the use of production libraries?

7.4 Quality assurance monitoring

Because of the more rapid development of applications, it is necessary for an organisation to monitor and control the quality of the code produced. The minimum requirements are the ability to determine what applications exist, what the interactions between routines and applications are, and who 'owns' them. Other requirements may be facilities to monitor, and possibly prevent, the use of particular constructs. Few AGEs address this area.

Does the AGE incorporate any facilities that can be used for Quality Assurance monitoring, for example utilities which scan or monitor applications and produce a statistical profile of where resources are used?

Can an application cross reference report be produced?

What controls are available to restrict use of particular AGE facilities to specific users?

Standards conformance	Increased concern with quality control has led to a desire to meet national and international standards for quality. In particular there is a demand to meet BS 5750 and ISO 9001. The three main characteristics of the former are the ability to record and monitor prevention, appraisal and failure information as a means to monitoring the quality of a product during its development cycle.

What facilities are provided to record prevention, appraisal and failure data?

Are any other facilities provided in order to assist in the development meeting BS 5750 standards?

Are there any similar facilities to help in adhering to ISO 9001?

7.5 Maintenance

Maintenance, in respect of the detection and correction of errors, is obviously a significant factor in quality control. Largely, the facilities which relate to this element of the maintenance function, have been covered under the heading of debugging in Chapter 2, and version control and configuration management above.

Are there any additional facilities provided, which assist in the maintenance function, as it relates to the correction of errors?

7.6 Performance monitoring and control

Some AGEs contain very powerful file processing constructs. Used carefully they can be extremely effective; used wrongly, they can be very inefficient. The AGE should contain facilities for gathering statistics on the use of such constructs and on the resources they consume. It is also desirable to be able to impose limits on the resources available to particular user authorisation levels, in order to prevent inefficiency.

A major benefit of AGEs, particularly when used with Relational DBMSs, is that they permit the end user to formulate his own ad hoc enquiries. Uncontrolled use of such a capability can, however, lead to serious performance problems.

Does the AG include accounting constructs to record the usage of the software and/or the resources used by the software?

Most DBMSs attempt to satisfy queries using some form of optimiser to perform the action in the most efficient manner. Some DBMSs will attempt to answer queries irrespective of the amount of resources used. What is required therefore is some form of governor that will estimate the amount of resource to be consumed by the transaction and will prevent the transaction being started or completed if this is deemed desirable. This was discussed in Chapter 5.

7.7 Effect on the organisation

The introduction of some products into a department is likely to have a much more significant impact on the way that department works, than will other products. 'Tactical' products are those which can easily be introduced to solve immediate one-off problems; they will involve a short educational and installation lead time. 'Strategic' products require a much greater commitment, in order to get the correct infrastructure into place, to use them effectively.

Strategic products also usually imply a commitment to using a true 'database' approach and to the use of a DBMS; although this would not necessarily be true for an OO throughout AGE, which could itself be classed as strategic whether it was integrated with a DBMS or not. Some products could be used for either.

Would you describe your AGE as being for strategic or for tactical use?

What form of organisational infrastructure is recommended to make optimal use of the AGE?

Does the product require a centralised data dictionary?

Does the AGE documentation recommend the appointment of a Database Administrator (DBA)?

Is the AGE part of an integrated set of products, for example DBMS, Data Dictionary, Office Automation etc?

Can the AGE be used to develop and run completely separate applications or are all applications combined into a system at some level (perhaps due to the specification being held in a single library, or possibly because the run time code needs to run in a shared environment)?

Does the AGE only interface to a fully functional DBMS?

8 Environment independence

Environment independence is important if it is likely that the application will be moved to another hardware or system software environment. An advantage of the AGE approach is that the aspects of the application which are dependent on the technical environment, are separated from the specification, which should allow the specification to be implemented on other environments with minimal changes.

This chapter refers both to the environment in which the generator operates and on which the resultant applications may operate. These are not necessarily the same.

The generator component consists of two parts: specifying the application and 'compiling' it. It is possible that these two could have different environments, but unlikely.

8.1 Hardware independence

If the AGE is to be used in multiple environments or in a distributed environment, then it will be undesirable if it is dependent on a particular hardware architecture. This will be less critical if that architecture is supported by multiple vendors (for example, Unix-based RISC [Reduced Instruction Set Computing] systems) rather than just a single supplier.

Is the product specific to a particular hardware architecture? Is it specific to a particular vendor's equipment?

To what extent will this impact on the environments within which a generated application will run?

Facilities available will be terminal dependent.

What type of terminals does the AGE support (for example synchronous, asynchronous, character, bit mapped, colour etc)?

	Portability	It may well be that the same AGE is intended for use on a number of projects which will be developed in diverse parts of an organisation. This may require the ability to operate on a wide range of hardware platforms with divergent operating systems. This will often be the case for distributed systems.

For what hardware platforms and associated operating systems is the AGE supported?

Are any of the versions of the AGE cut down in any way, in order to run on smaller systems?

If so, to what degree?

It is commonly the case that vendors will claim Unix support as a catch-all term. Even where particular hardware platforms run the particular version of Unix named, this does not guarantee that the AGE will run on that hardware unless a specific port has already been achieved. The length of time to do a port to a new platform will very considerably.

How long does it take, on average, to complete a Unix port of the AGE, to a new platform?

Provide user references for previous ports. |
| | Standards conformance | The easiest way to ensure portability is to insist on standards. These apply at various levels. Depending on the hardware environment these may include OSI, Posix, X/Open, X/Windows and so forth.

To what standards does the AGE adhere?

What user interface standards are supported? |
| 8.2 | **Software independence** | In some respects one cannot usefully discuss hardware independence without referring to the operating |

Chapter 8
Environment independence

system. This we have just done. In others, the operating system can be taken on its own.

As was mentioned above, it is sometimes the case that different variants of the product are available on different machines. In such cases, applications may not be portable. This may be reflected in some of the other answers.

If the application generator operates on a range of machines or under a range of operating and TP systems, at what level is portability achieved?

Generators which require batch facilities are usually less effective in a prototyping environment. Some vendors appear to try to hide the fact that a batch process is necessary as part of the generation process.

Do any components of the product require a batch environment? If so, which?

Application generation frequently makes full use of interactive development. It is useful to be able to do some of the application specification/generation in a batch mode if online terminals are scarce.

Can those components of the product which are normally used in an online environment also operate in a batch environment?

Required operating system modifications may cause support or maintenance problems.

Is it necessary to incorporate any non standard features within the operating system or the TP monitor to support the applications generator?

Interoperability

Whereas portability reflects the ability of software to run on different hardware platforms and under different operating systems, interoperability refers to the ability to interact with other software products. Generally speaking the ability to interoperate will be dependent on the standards (de facto or otherwise) adopted, both for the AGE and generated applications. Conformance to standards for the various types of product are addressed below.

What specific third party products will the AGE interoperate with, other than those specifically defined by their adherence to standards?

What facilities does the AGE incorporate to ensure that generated applications conform to selected standards?

Standards Conformance

The degree to which the AGE will interoperate with other-vendor products, by means of gateways, shared interfaces to dictionaries and so forth, has already been discussed. This applies to DBMS (SQL), TP Monitor (XA compliance with the database) and CASE tools (SSADM), with particular reference to the data dictionary.

In a distributed environment the one standard that has not been discussed, and which will help to facilitate interoperation, is the OSF's Distributed Computing Environment (DCE).

Does the AGE support, or are there vendor plans to support, the OSF's DCE?

If so, to what extent and how will this impact on interoperability?

9 End user interface

The usability of the application is likely to be a major factor in determining the success of the implementation. While the look and feel of the application is largely dependent on the skills of the development team, they are constrained by the functionality of the product. The criteria below concern the environment that can be built for the end user. Note that some AGEs operate in more than one environment, for example on bitmapped workstations as well as character based terminals.

It is normally true that providing a high quality user interface, perhaps using colour, or a bit mapped workstation, is likely to increase the cost of the project. It may often be the case, however, that the usability of the interface is what determines the ultimate success or failure of the project. Such factors should not be neglected, even at an early stage of a project.

9.1 Invocation

The user should be able to invoke the required application as easily as possible and with a minimum of keystrokes. The ideal is for the user logon with password to lead automatically into the application main menu.

Some systems do not differentiate between using the product to specify an application and using it to run an application. This is not the best approach when attempting to construct 'black box' application systems. Also, it is often advantageous to be able to merge the transaction processing system's logon procedure with the application's logon procedures to avoid the user having to sign on twice.

To use a generated on-line application, does the user:

- *invoke the same logon sequence as used by the generator when specifying an application?*

- *invoke a logon sequence designed as part of the application?*

- *use the transaction processing monitor's standard logon sequence?*

- *other? (please specify).*

9.2 Navigation

Simple and consistent mechanisms should be available to enable the user to move from one field to another, or one screen to another.

What is the control mechanism to move from one screen (exchange) to the next?

- *the user specifies the transaction code of the next screen (exchange) explicitly*

- *the user is automatically returned to a central menu where he specifically selects the next exchange*

- *the application decides based upon the user's input*

- *the user chooses a course of action from a list of options (e.g. next screen, previous screen, exit etc).*

Function or control keys are used to switch states within an application and are therefore useful. However, their inclusion may depend upon the terminal's characteristics.

Are function keys supported? Are these programmer definable?

Are function keys consistent across different systems and terminal types?

Can the developer change the keystrokes required for actions such as accept, abort etc; plus the terminology used by the AGE supplied form menus, for example Add, Update, Query etc?

Chapter 9
End user interface

9.3 Dialogue

The application developer will design a dialogue for use by the end user. This dialogue will (normally and preferably) consist of a hierarchy of menus and forms. The capability of the workstation usable by the product, and affordable by the project, dictates the type of environment that can be built for the end user.

End users are becoming increasingly familiar with sophisticated interfaces and are beginning to demand the same functionality from an AGE.

Are bitmapped workstations with mice supported?

Are pop-up or pull-down menus available?

What types of menu selection are available?

- *selection by mouse*
- *ring menu*
- *numeric designation*
- *alphabetic designation*
- *first character*
- *scroll and highlight using cursor keys*
- *other.*

Are windows supported?

9.4 Presentation and rendition

The appearance of what is seen on the screen is limited by the terminal or workstation. The application can be made much easier to use by careful design, including using the display rendition capabilities to differentiate between the different categories of information seen on the screen. The availability of graphics to permit lines, boxes, circles etc, to be drawn on the screen is also beneficial.

Graphical user interface	If an application is developed in a GUI based AGE, then it is usual that at least some of the facilities that are intrinsic to the GUI may be incorporated into applications. Since this is as much a factor in development functionality as it is in end-user presentation, this topic was covered in detail in Chapter 3.
Other facilities	A range of different display capabilities will be required for general applications, but the selection may be constrained by terminal characteristics.

Can the following field attributes be specified within the AGE?

- *protected*

- *hidden (not visible - used for passwords, etc)*

- *high brightness*

- *reverse video*

- *colour*

- *blinking*

- *font*

- *character size*

- *other (please specify).*

The availability of graphics enables the developer to build systems that can provide a familiar image to the user, for example by emulating existing manual forms.

Can the screen painter paint graphics, or at least graphics characters?

Chapter 9
End user interface

9.5 Variants

Often one dialogue can be made to suit the requirements of more than one user role. This would be where their requirements only differ slightly; perhaps some menu options are not available to one role, or perhaps the existence of certain fields are not disclosed.

One approach is to allow variant sets of forms to be retained with an indicator to link them to particular logons. Object oriented inheritance can also be useful in this respect.

Can variants of the dialogue and forms be made available to different users or groups of users? How can this be achieved?

9.6 End user help system

Help systems reduce the training overhead and can be an essential learning aid for beginners, or occasional users of a facility. Help systems can either be full blown parallel applications, or can be simply pop-up windows which appear on request.

Can help screens be programmed?

Can the help system be generated automatically from the data dictionary?

Is help context sensitive?

Can the help system be browsed?

To what extent can the developer control how the user obtains help?

Is there a limit on help available at any level? If so what are the limits

9.7 Session concurrency

It is beneficial if the user can suspend one activity and invoke another, perhaps for cross reference purposes. This could be achieved by windowing, by multiple sessions, or by split screen working. Users are increasingly demanding this functionality - for example to answer telephone enquiries.

Can the user run more than one application at once?

Can the second application be a dissimilar application, for example a word processor or mail system?

9.8   Error messages   Error messages should be clear, concise and meaningful; cryptic error messages do little to inspire user confidence.

Can the developer supply text to clarify the error messages emanating from the AGE, the DBMS or from the operating system?

Is the error code supplied in this way actioned by:

- *the operating system?*

- *the DBMS?*

- *being held and used by AGE?*

- *the dictionary?*

- *other? (please specify).*

9.9   Skill levels   The requirements of experienced users are not the same as those of novice or occasional users. It may be beneficial to have 'short cut' or 'expert commands' for skilled users.

Does the system support the concept of different levels of user skill?

Can expert, 'short cut', commands be implemented?

Chapter 10
Security

10 Security

Security of data, and of the application development tools that access it, is of vital importance to many projects. Use of the AGE may bypass many of the underlying DBMS controls, hence control at the AGE level is essential.

10.1 Access control

To what extent will the developed application allow different users to access only the functions to which they are entitled.

How can users be restricted to particular parts of the application, or from accessing particular datasets, tables or fields?

Does the AGE rely on DBMS access controls?

Will the AGE support access control devices such as badge readers etc?

Can users be restricted to particular tools or subsets of tools within the application development environment?

10.2 Encryption

It may be important, especially if using a distributed system, to be able to provide encryption facilities. These fall within two categories.

Data encryption

Data encryption will prevent unauthorised use of information. It may, however, cause a significant performance loss.

Can the AGE encrypt data or is it dependent on DBMS or Network Operating System capabilities?

Password encryption

A less onerous form of encryption, in terms of its affect on performance, is the use of encryption for passwords only.

Can the AGE encrypt passwords or is it dependent upon other system software elements?

10.3 Other issues As security becomes increasingly important, issues other than mere provision of passwords, access levels and encryption become necessary. These are not typically within the scope of an AGE and are mentioned here only for the sake of completeness.

Virus protection Viruses are increasingly common. Fortunately, anti-virus software is becoming ever more sophisticated. It should be standard practice to run suitable programs against any new software to be loaded.

Unauthorised versions In certain environments it is essential that all users of an application use the same version of software. If this is the case, then either internal procedures for the issuance and use of new software need to be in place or, in a distributed system, it may be possible to use a software distribution package.

Unlicensed software It is important that organisations do not breach vendor agreements by utilising unlicensed software, or using it in ways that break the licence agreement. Suitable software packages are available for use in a distributed environment, in order to control this. Some packages also include unauthorised version control (see above).

Network issues In a Local Area Network, or other distributed system, security can be altogether more of a problem than with a single processor and dumb terminals. For example, it is very much easier to put in an illegal tap. Access security and password encryption should be considered minimum requirements where there is any sort of confidential information available. Additional security features include time and place log-on security. This was addressed in Chapter 3.

Chapter 10
Security

Standards conformance

The main work on standards for security has been conducted by the US Department of Defense, who's results have been published in the so-called Orange Book. This lists seven levels of security from D (the lowest) to A (the highest). Few products have yet incorporated these standards, particularly at the lower end of the market, though a number of vendors have announced intentions with regard to C2, with promises of B1 to come.

Is the AGE certified to any Orange Book standard?

Does the vendor have plans to apply for certification?

If so, at what level?

Does the AGE provide facilities that would enable applications to meet Orange Book standards? If so, please describe.

An area in which there are emerging de facto standards is in encryption algorithms. The best of these (for example, DES - Data Encryption Standard) have been developed in the United States, and are under strict export control.

If an encryption algorithm is provided, what methodology does it use?

Is this controlled for use outside the USA? If so, how can the vendor assist in gaining usage rights?

A further area in which a de facto standard appears to be arising, is in the Kerberos authentication server for distributed processing. This forms a part of the OSF's DCE. Using Kerberos, a user wishing access to a server, first applies to the Kerberos server which validates the request, before allowing the client to address the data server in question. Some vendors have gone further by providing their own extensions by way of verifying the services to which that client is entitled from that particular server.

Will the AGE run with the Kerberos authentication service?

What facilities does the AGE provide, in order to enable generated applications to run with a Kerberos server?

11 Vendor and product credibility

Application generation environments are frequently the subject of 'imaginative' marketing. Some AGEs are produced by small or relatively unknown software houses or are written and supported in foreign countries and marketed here by agents. Other AGEs are new to the marketplace and are therefore as yet untried. Such AGEs are not necessarily of poor quality, but it is necessary to assess the likelihood of the software and the marketing agency still being viable in the future, before committing to using the product, irrespective of its technical merit.

11.1 Quality of product

Application generation enviornments should be constructed to a high quality using rigorous, structured development and testing techniques.

Is the product specified using a formal definition language or method such as VDM or Z?

Does the product developer subscribe to and comply with the requirements of British Standard (BS) 5750 or equivalent documents? (This deals with a supplier's capabilities to operate a quality management system in the design, manufacture, installation, inspection and testing of a product).

Does the product developer have certification to BS5750 or equivalent? If not, why not??

Has the product been submitted to any independent authority (eg the National Computing Centre Ltd (NCC)) for evaluation or certification/validation? If so, are the results available? Are any independently reached performance figures available from such authorities?

What guarantees are there against defects in the product?

11.2 Product development status

Information on the development status of the product is essential before making a long term commitment to its use.

What is the current development stage of the product, for example:

- *static?*

- *stable but in the process of being cosmetically enhanced (ie minor changes and improvements in presentation, or the way(s) in which the product interacts with the user, are in preparation. Any such changes will not affect basic, user functions)?*

- *being functionally enhanced?*

- *in the process of being developed for use on other machines?*

Have any enhancements been introduced recently? Are there any which are under development? If so please list planned enhancements and the target dates for the introduction of such enhancements.

When was the last major, new version (as defined by the supplier) released and when is the next major, new version planned for release?

How does the company determine when a system requires enhancement and the nature of the additional or supplementary facilities and features which are to be incorporated?

Are there any weaknesses which have been identified in the current version of the product?

What plans are there for the product over next 3 to 5 years? Will the product be different to today's version? If so, what differences will there be?

How is system updating arranged to take account of new developments and legislation?

Have any overseas products been localised (eg date format, currency symbol)?

How many updates have been issued in the last year?

How many errors and corrections are outstanding?

11.3 Vendor assessment

The supplier assessment will take into account the size of the supplier, whether they are the originators of the software or simply agents, how long they have been producing or marketing software, and whether they are a company based in Britain, Europe or elsewhere. The term supplier should be used in the widest sense, ie where the supplier is not the developer, all organisations involved in the development, marketing and support of the AGE should be investigated.

Some AGEs are produced originally by small independent software houses and then marketed, sometimes under another name, by computer manufacturers. This section should help to identify such products. Also, products simply marketed rather than developed by a supplier, are likely to enjoy a lower level of on-going support.

Name, address and telephone number of supplier.

Name(s), position(s), address(es) and telephone number(s) of the person(s) to contact for further details, if necessary, regarding

- *Marketing information*
- *Technical support.*

How long has the supplier been in operation

- *in the UK?*
- *world-wide?*

Was the product originally developed by the above supplier?

What organisations have used the supplier's services in the past?

Does the supplier have a range of products covering related topics, ie is it an area in which he specialises?

Is the supplier a subsidiary of any other company? If so, please give details.

How many years has the supplier been active in the development and/or marketing of application generation environments?

During the last year, what percentage of total revenue has been derived from AGEs?

What percentage of total profit or income has been contributed to research and development of AGEs? (ie future facilities, CASE, etc)

How many employees are dedicated to the development of AGEs?

11.4 Product background

Most application generation environments start their life in a slightly unstable state; some never achieve stability. If an AGE has a reasonable number of production field sites (not simply copies out for approval or copies distributed but not seriously used), then the product's capabilities and potential may be assumed to be at least adequate and the risk involved in selecting such a product is less than that of a new and untried package. New releases of a product may however be potentially risky.

Development ancestry

Potential buyers should establish when the 'product' was first available rather than the concept. Some AGEs are developments of tools used internally by the supplier. Sometimes these early internal versions are quoted to imply that the product has a better 'history' than is the case.

Chapter 11
Vendor and product credibility

When was the AGE first installed at a customer site for customer usage?

What is the source and history of product(s) under consideration?

For how long has the AGE been commercially available?

What was the development environment (ie the machine on which the AGE was developed in the supplier's organisation)?

Did the supplier write the AGE, or is he acting as agent?

Where is the software originator based, eg local, UK, Europe, America?

Development profile — Many AGEs are still in a state of development and enhancement. New features, facilities and environments are being added. While this may provide many useful new features it may cause problems if releases with desirable new features appear during a development.

Note: new versions of application generation environments usually incorporate either significant improvements in functionality or in performance. New versions are typically released on an 18 to 24 month cycle. Intermediate releases tend towards fixing bugs only.

How often are major product versions released? When was the last one released and when is the next one planned?

What enhancements, if any, are planned for the AGE and when will they be introduced?

What is supplier policy towards compatibility between versions?

How does the AGE supplier determine when a system requires enhancement and the nature of the additional/supplementary facilities which are to be incorporated?

Product usage

An indication of the numbers of users of the AGE, the sales profile of the AGE and other pieces of information such as product appraisal and evaluation reports can give a valuable insight.

How many user sites of the AGE are there

- *in the UK?*
- *within Government?*
- *outside Government?*
- *elsewhere in Europe?*
- *elsewhere?*

How many systems of this type have been sold in the UK (and worldwide) during the past 12 months?

How many existing users are there (particularly any within government) and what is their volume of transactions (ie the number of applications currently in existence and use)?

For how long have earlier, or original, users stayed with the AGE; or are all users (comparatively) recent?

Which is the nearest competing product available in the marketplace?

Describe any previous projects using this AGE with which the supplier has been involved both inside and outside of government. Please indicate the size and complexity of the jobs in broad terms.

Please give the name and address of reference site(s) which may be contacted if necessary.

Can other users be contacted, ideally in the same business area?

Chapter 11
Vendor and product credibility

When was the first system of the type being considered (or proposed) successfully installed at a customer's site? Please provide details of the site's location and others, if available.

Does a user group exist for the product in question? If so, please state:

- *whether it was formed independently of the supplier's organisation*
- *how long such a group has been in operation*
- *how active is it*
- *the number of active members*
- *joining/ membership fees*
- *the number of meetings held each year*
- *when and where meetings are held?*
- *the name, address and telephone number of the group's secretary.*

How closely does the supplier collaborate with any such user groups which might be established? Does the group develop wish lists of future enhancements, and how are these acted upon by the supplier?

Product information
Other sources of information other than those suggested by the product supplier can be valuable.

Are there any independent reports and evaluations on the AGE(s) being considered? Can copies of any reports be made available (IF NOT, WHY NOT?)?

User Profile
While many AGEs are usable by a wide class of users, most products are best suited to particular user profiles. Best results will always be attained when the correct tools have been chosen for the particular user

profile. Most suppliers will claim usability by all classes of user.

Who are expected to be the principal users of the AGE?

- *non IS staff*
- *analysts*
- *novice programmers*
- *experienced programmers*
- *others (please specify).*

11.5 Documentation

Application generation environments require adequate documentation. Frequently AGEs at the beginning of their life, or AGEs marketed by small organisations, appear with inadequate documentation. Other AGEs appear with large amounts of poorly structured documentation.

What information is available about the AGE before purchase?

What documentation is available, and how well is it presented?

What manuals and other documentation are provided when the AGE is purchased?

What other 'optional' manuals are available?

Can the documentation be copied by the user for his own use only?

What is the target audience for each manual eg management overview, system designer, application programmer, operator, etc?

Are the manuals available online?

What information is available on the technical content of the system, eg:

- *record formats*
- *database structure*
- *parameter tables*
- *validation mechanisms*
- *source code?*

In the event of the supplier going out of business, what arrangements are there for access to the source code, eg is a copy of the source code lodged with an Escrow Agent?

Do customers use the documentation provided or is there a need to develop instructions which are specific to each installation?

11.6 Training

Application generation environment packages in particular, usually claim to require relatively small amounts of training, but this is not always the case, and certainly won't be if they are OO throughout. Poor quality training will predispose staff against good products and may therefore affect a project's overall success. Note that length of training required is not a sufficient guide as this will depend on the complexity of the product.

Note that AGEs suitable for end user use may require separate introductory courses for programmers and non programmers.

Please state whether training is included in the purchase price of the proposed software system and:

- *where such training is normally carried out*
- *whether onsite courses can be arranged*

- *the nature and amount of training normally required to operate and use the AGE (based on the supplier's previous experience)*

- *the duration of training courses*

- *the individuals at which such training is aimed.*

Are appropriate training courses provided by the supplier?

Do any third parties offer training in the use of the AGE?

How much computing expertise is required by attendees?

Please describe any additional training related to the efficient and effective use of the system including details of cost, location, duration and frequency.

11.7 Support

Support will be required, especially when an AGE is first introduced and before the organisation has built up its own inhouse expertise. The type and level of support available will depend on the size of the supplier organisation and the number of sites they are supporting. There have been a number of instances where AGEs have enjoyed rapid market success but this has resulted in their support services being thinly stretched or staffed by poorly qualified personnel. Support quality is also likely to be dependent upon where software development is done. If all development is done overseas, then the local knowledge of the internals of the software is likely to be reduced and the time taken to fix bugs increased.

Some AGEs are purchased or marketed by UK suppliers, but not written by them. Where this is the case the level of UK support may be found wanting for newly established AGEs.

General

Where and by whom is support undertaken?

Where are the support services located?

Chapter 11
Vendor and product credibility

What is the policy for supporting previous releases of the AGE and how many versions are currently supported?

To what extent is modification by users allowed without affecting support?

How are queries and problems dealt with after installation?

For which aspects of the implementation will the supplier be responsible (eg hardware and software installation, system and data conversion, user training)?

Pre-sales

Is a demonstration available?

What are the arrangements for a trial of the AGE?

Does the right exist to reject the product if it fails user specified acceptance tests?

Will any verbal claims and promises made by sales people be written into the standard contract?

Who will provide support/answer queries, and how accessible are they, eg by telephone, office hours only?

Installation

What maintenance and support services are available during installation of the AGE?

Will specific personnel be allocated to this project (full or part time; at the beginning of, during and after implementation)?

Type and level

What maintenance and support services are available once the AGE is operational?

How many technical support staff are supporting how many users?

How many technical support staff are available in the UK?

Does the supplier operate a 'hot-line' service for urgent user enquiries and fault reporting? If so, is the service part of a system maintenance agreement and please state:

- *the average response time*
- *the longest response time*
- *the way in which the system operates.*

How long does it take for a supplier's hot-line to answer, and how long to resolve queries?

Is there a charge for hot-line support?

What procedures are available for reporting problems and what action and priorities are assigned to rectifying faults?

Describe the circumstances in which onsite maintenance/ assistance would be given. Would such services be provided by a sales representative or by a software expert/engineer? What would be the contractual response time for a call for assistance?

Fault correction

Are details of system faults and required corrections circulated regularly to users? Is the software supplied with all corrections applied or are the corrections (fixes) supplied separately for incorporation by the user?

How are faults corrected (for example, by means of a new software issue, letter of notification, onsite assistance; or by telephone contact)?

Are new versions of the AGE automatically sent to users?

What are the escalation procedures for fault correction? When will the Managing Director become aware of a serious fault?

11.8 Enhancements

The methods and procedures by which enhancements to software products are handled are extremely important. They should be designed to minimise or avoid any disruption.

Chapter 11
Vendor and product credibility

What arrangements can be made for future changes which may be required by the user?

What are the arrangements for future changes in requirements, and how will the work be costed?

How upward compatible is the AGE for changes to

- *the hardware?*
- *the operating system?*

Are there facilities for users to 'customise' the AGE?

11.9 Related products

Some AGEs have 'core' or 'shell' application systems available; these could save development time

Does the AGE supplier market application products eg

- *personnel system?*
- *financial systems (eg general ledger)?*
- *others? (please specify)*

How are these products supported?

Appraisal and Evaluation Library
Application Generation Environments Volume

12 Project specific requirements

Any other requirements specific to the Departmental IS Strategy and/or project but not fully covered in other parts of this volume.

Appraisal and Evaluation Library
Application Generation Environments Volume

Chapter 13
Costs

13 Costs

It is usually a strategic objective to minimise costs but this is subject to meeting other requirements. Often costs are tangible but benefits are intangible. Within the CCTA IS Guides it is recommended that costs should be calculated over the life of a study or project, usually between 5 and 10 years. While cost comparison is done in detail for the final selection of a product from the short list of approved products, there is also a case for including costing in the higher level formulation of that short list. For this purpose (ie software comparison), costing need not be done at a detailed or absolute level; approximate relative costs are sufficient.

13.1 Software

Software costs include a basic license cost plus usually a recurrent annual maintenance charge. When prices are given it should be indicated whether these are inclusive or exclusive of VAT.

What is the basis of sale (ie lease, rental, purchase) and is the software the subject of copyright? Please provide details of licence fees which may be appropriate and the way(s) in which the licence operates, eg site licence, single system, total organisation? If a licence only covers the use of a package on a single, identified computer system, can it be run elsewhere in an emergency?

How much does it cost to buy the product outright? Does this cost include a copy of the source code?

Can the product be acquired on a trial basis and if so for how long and what are the costs involved? Are these costs discounted from any subsequent purchase price?

Does the product require any particular separately purchasable prerequisites or components (software or otherwise) (from any vendor source)?

How much does it cost to rent or lease the product

- *per month?*
- *per year?*

What are the minimum and maximum rental periods?

What are your terms for multiple copies of the product

- *on a single site?*
- *on multiple sites?*

For the following customer support services, please give details of those which are available and the associated costs that are not included in the basic software charge:

- *design and configuration planning*
- *preliminary planning support*
- *development of clerical and operational procedures*
- *data and system conversion*
- *implementation.*

If customisation of the package is available from the supplier, what is the basis for charges, and how many staff are available for such changes?

What happens if we buy your system but, after several months cannot achieve the productivity gains claimed? Has this happened to other purchasers?

What installation support is included within the purchase price? Does the price include the cost of new versions?

Is any warranty provided?

Is software support provided? If so, what is the cost?

Chapter 13
Costs

Development vs
Maintenance costs

In costing the application there are two considerations vis-a-vis the distinction between initial development and on-going maintenance. In the first instance is whether there are distinct licencing arrangements.

Is there a separate run-time, as opposed to development price, for the AGE?

If so, what are the criteria and figures involved?

More significant in overall terms will be the extent to which the AGE reduces the on-going maintenanace function. Independent surveys suggest that 80% of project development costs are in the maintenance stage. Therefore the extent to which this is minimised will greatly effect the overall project costing and, in consequence, the cost-justification in selecting one AGE rather than another. Unfortunately this can only be a subjective judgement. Certainly vendor estimates cannot be relied upon. It is likely, however, that this is an area in which object oriented AGEs will score heavily.

This is an area which should be borne firmly in mind when making any supplier cost comparisons.

13.2 Hardware

Hardware costs vary with different product sets. The existence of a TP monitor and of multi-threading software can result in significantly reduced CPU costs. If the product is to be installed on existing hardware, then enhancement of this hardware may be necessary.

As with software, hardware is likely to have an initial capital cost together with a recurrent maintenance cost. Costs may have to be considered for the system as a whole and not just the package element.

What is the cost of enhancement of existing hardware (for example additional disc and/or memory) in order to support the product?

13.3 System operation and maintenance

System associated costs vary with the type of development tools used. 4GL tools will reduce development time and probably maintenance costs but will consume more hardware resources during development and operation. 3GL tools may produce efficient systems but take a long time to develop and may be more expensive and difficult to maintain. Different costs may be incurred depending on the type of licence, for example development or runtime (as noted above). If purchasing the latter, ensure that it includes all necessary facilities.

With certain types of product, significant savings may be made by not purchasing unnecessary copies of development software (generators, editors, compilers, etc) in an environment where applications are run on multiple sites, but development is done centrally.

Please specify the scope and terms of the maintenance contract; and describe those services which are standard and those which are optional.

What are the maintenance/support costs and arrangements for corrections, upgrades and new releases?

Does the maintenance charge cover the issue of new software releases/versions?

13.4 People

People costs are affected by factors such as the number needed, training requirements and their commercial worth. Sophisticated development tools reduce the number of people required for application development and probably reduces their training costs. However, such trained people may be commercially attractive and require a high salary to retain them. Sophisticated tools may require skilled infrastructure support and such people are likely to be expensive. Take on of new technology may require outside consultancy support that, while usually cost effective, will be expensive.

What is the cost of any training not provided free of charge when the product is purchased?

Please state whether training is included in the purchase price of the proposed software system and:

- *where such training is normally carried out*
- *whether on-site courses can be arranged*
- *the nature and amount of training normally required (based on your company's previous experience)*
- *the duration of training courses*
- *the individuals at which such training is aimed.*

Please describe any additional training related to the efficient and effective use of the system including details of cost; location; duration and frequency.

Is consultancy support available from the vendor? If so, what is the cost?

13.5 Documentation

Normally a package should be supplied with all relevant documentation necessary to operate it in an efficient and effective manner. Sadly, some product's supplied documentation does not totally fulfil these requirements and additional documentation may need to be acquired. Some suppliers, as part of their commercial pricing policy, may choose to charge separately for appropriate documentation.

What is the cost of any manuals not provided free of charge when the product is purchased?

Can manuals, once purchased, be copied for use only by the purchaser?

Appraisal and Evaluation Library
Application Generation Environments Volume

Annex A
Criteria hierarchy

A Criteria hierarchy

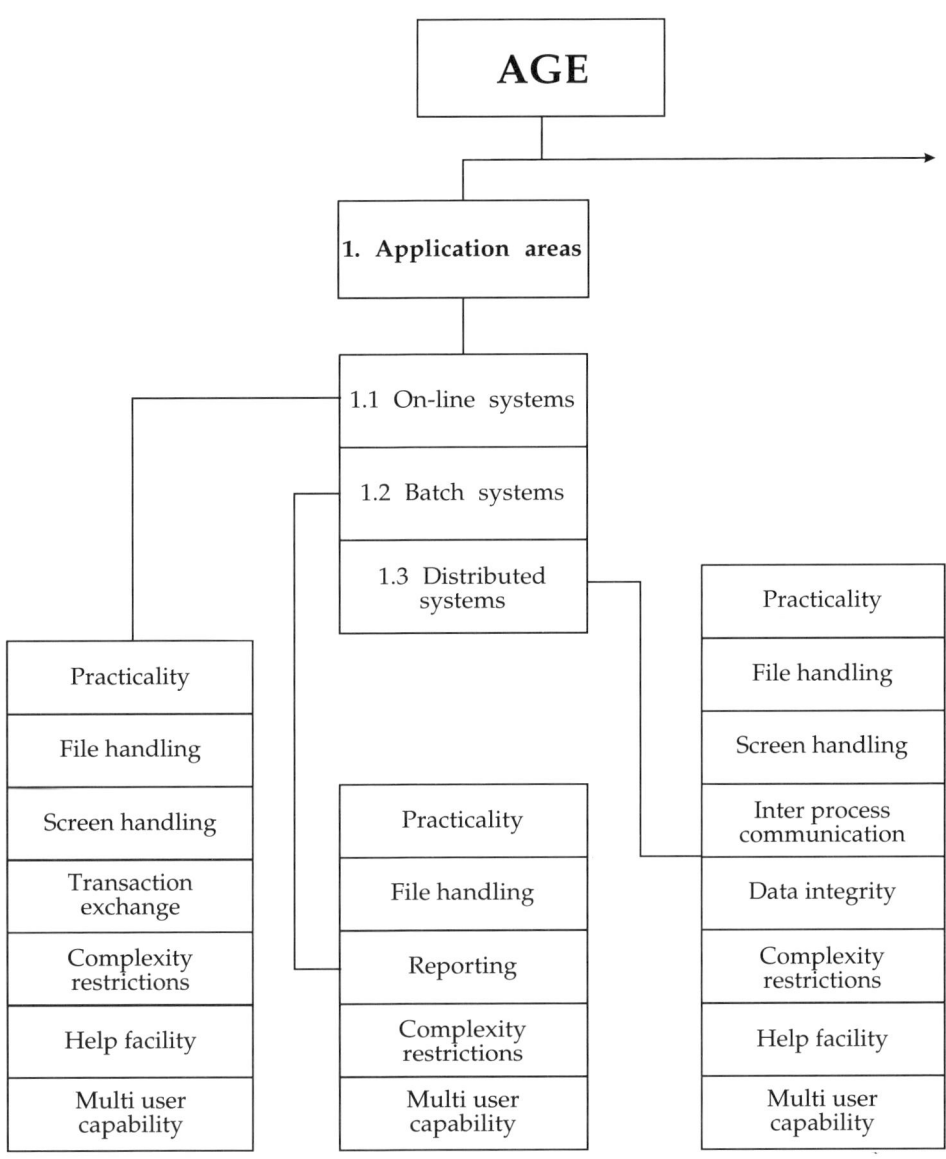

137

Appraisal and Evaluation Library
Application Generation Environments Volume

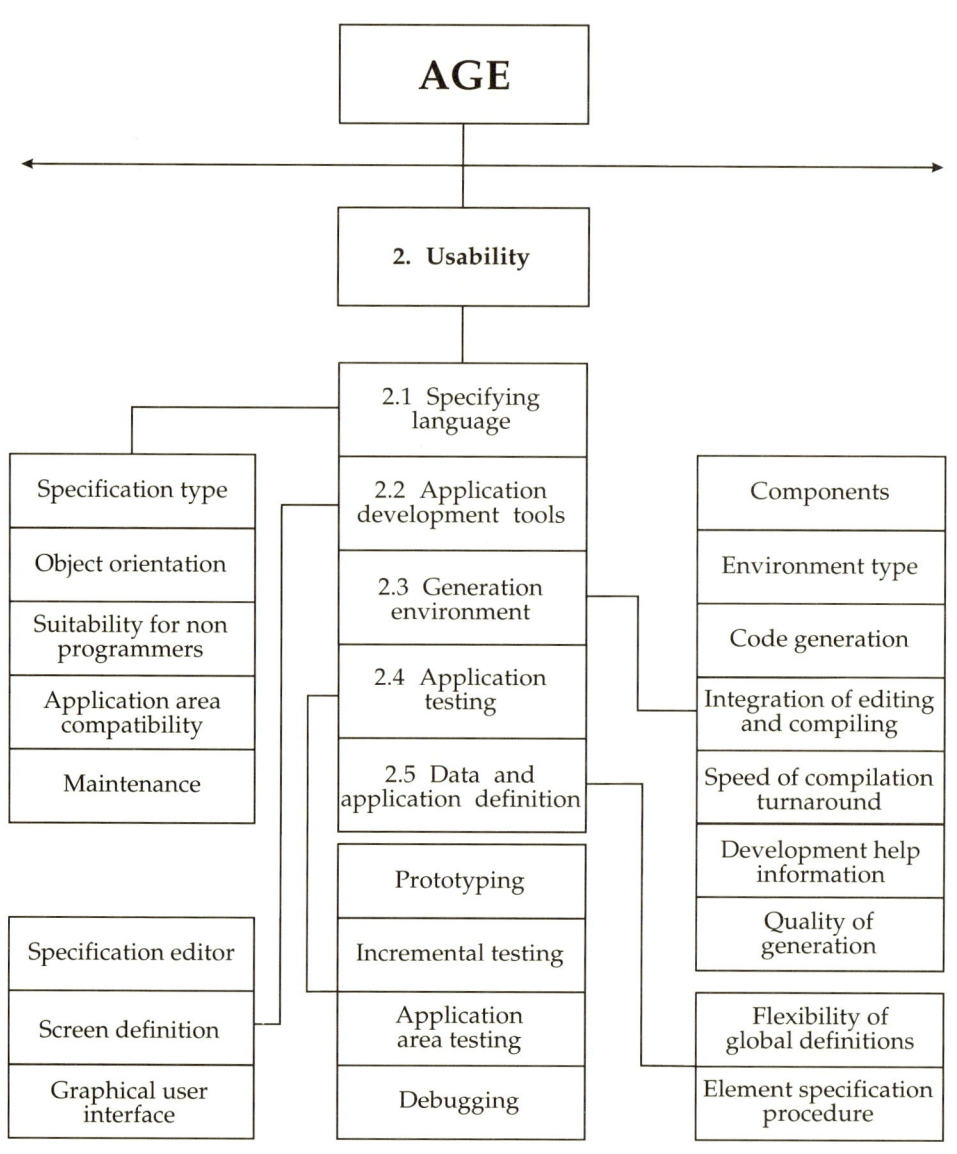

Annex A
Criteria hierarchy

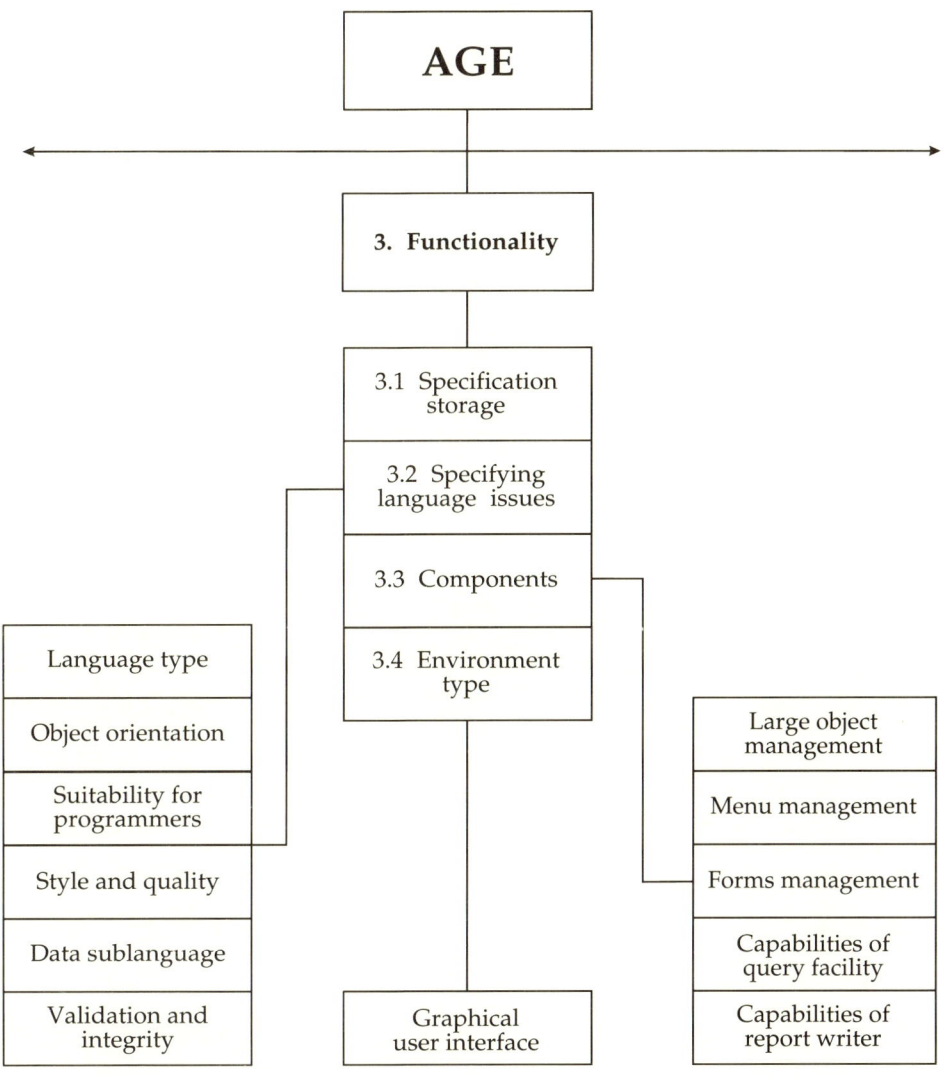

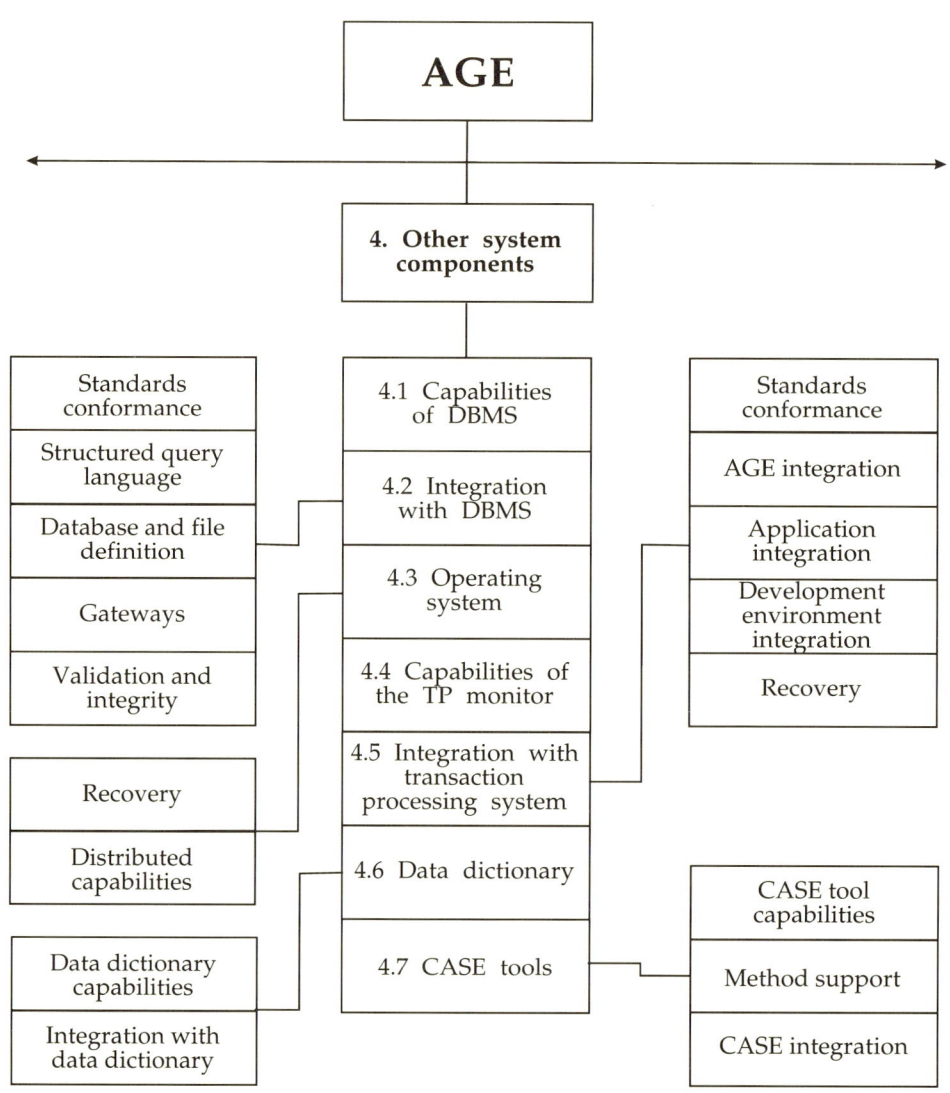

Annex A
Criteria hierarchy

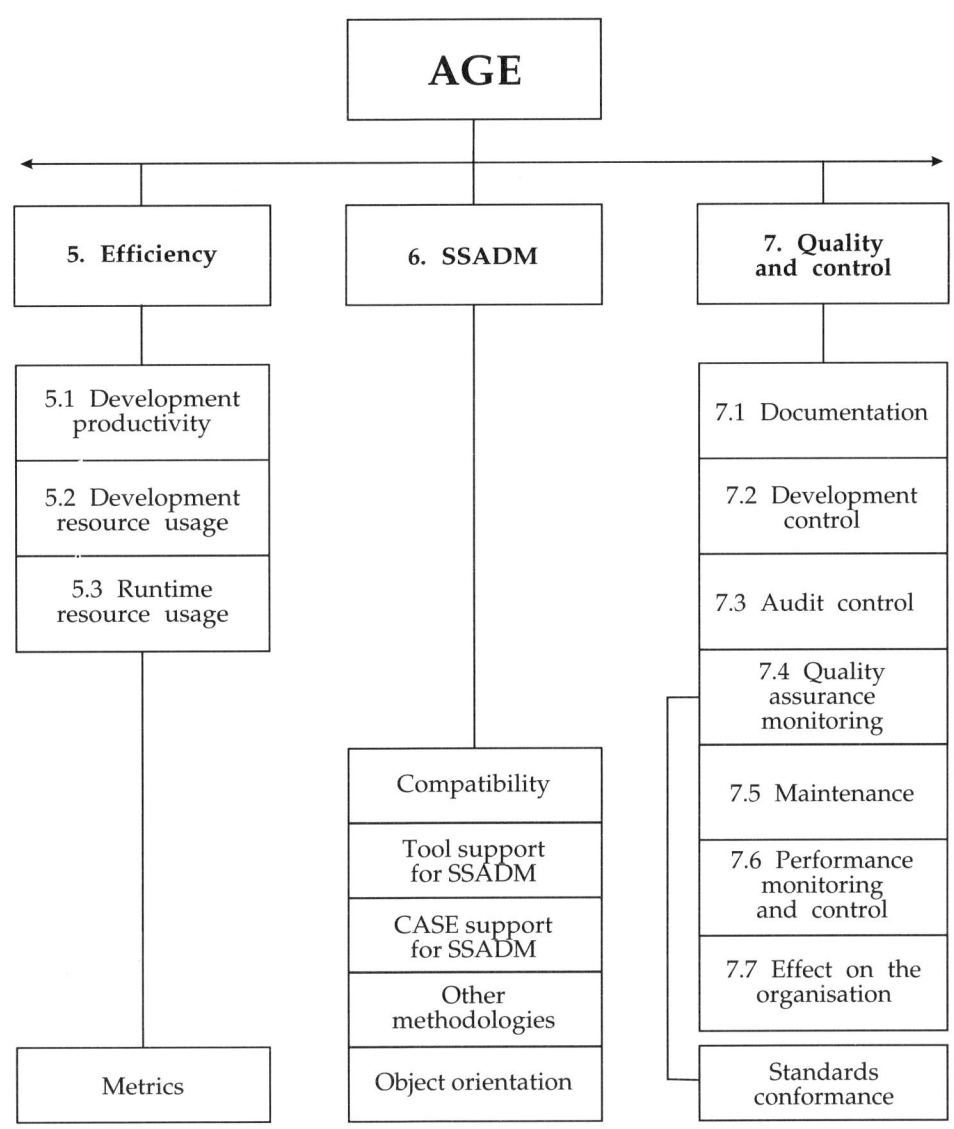

Appraisal and Evaluation Library
Application Generation Environments Volume

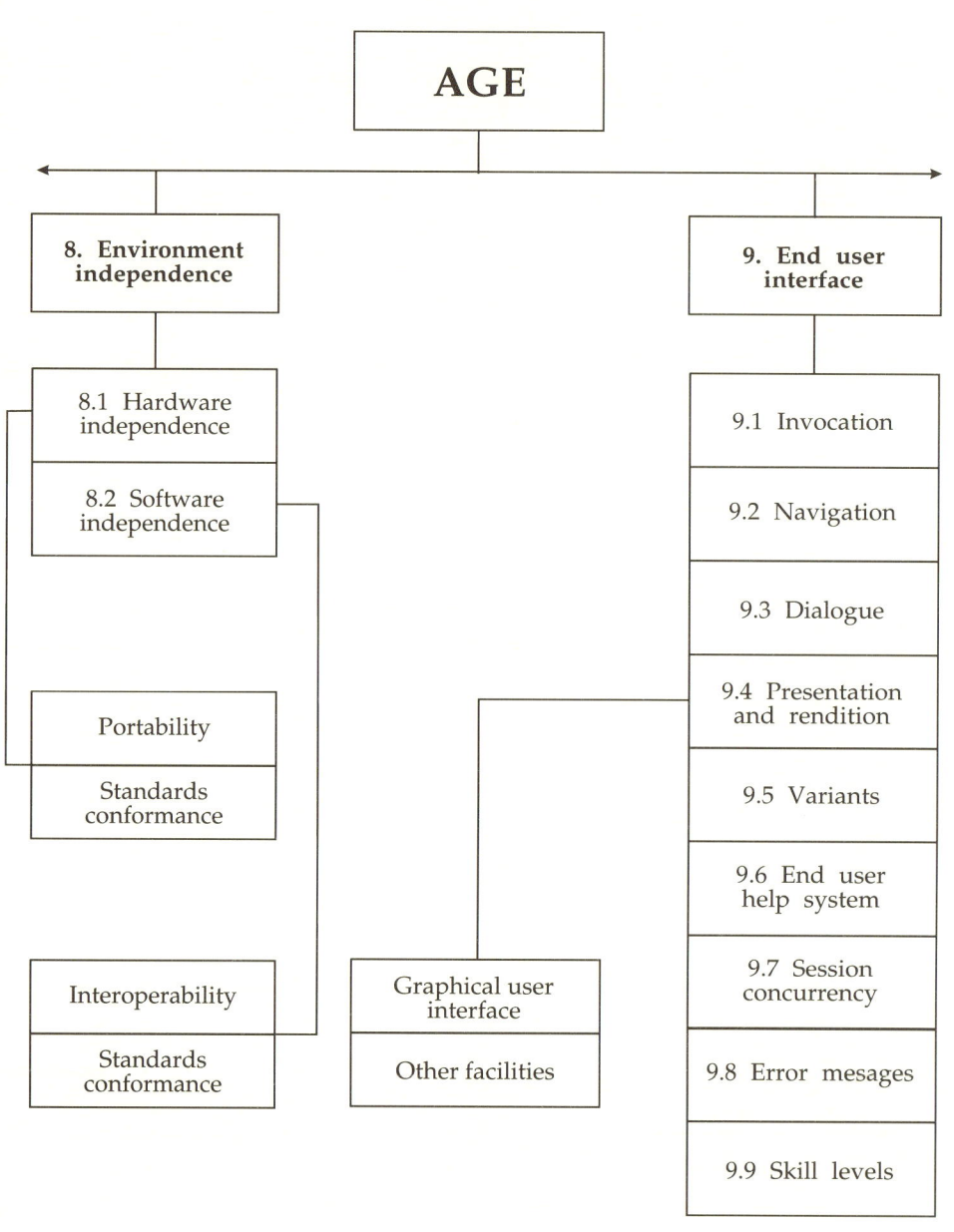

Annex A
Criteria hierarchy

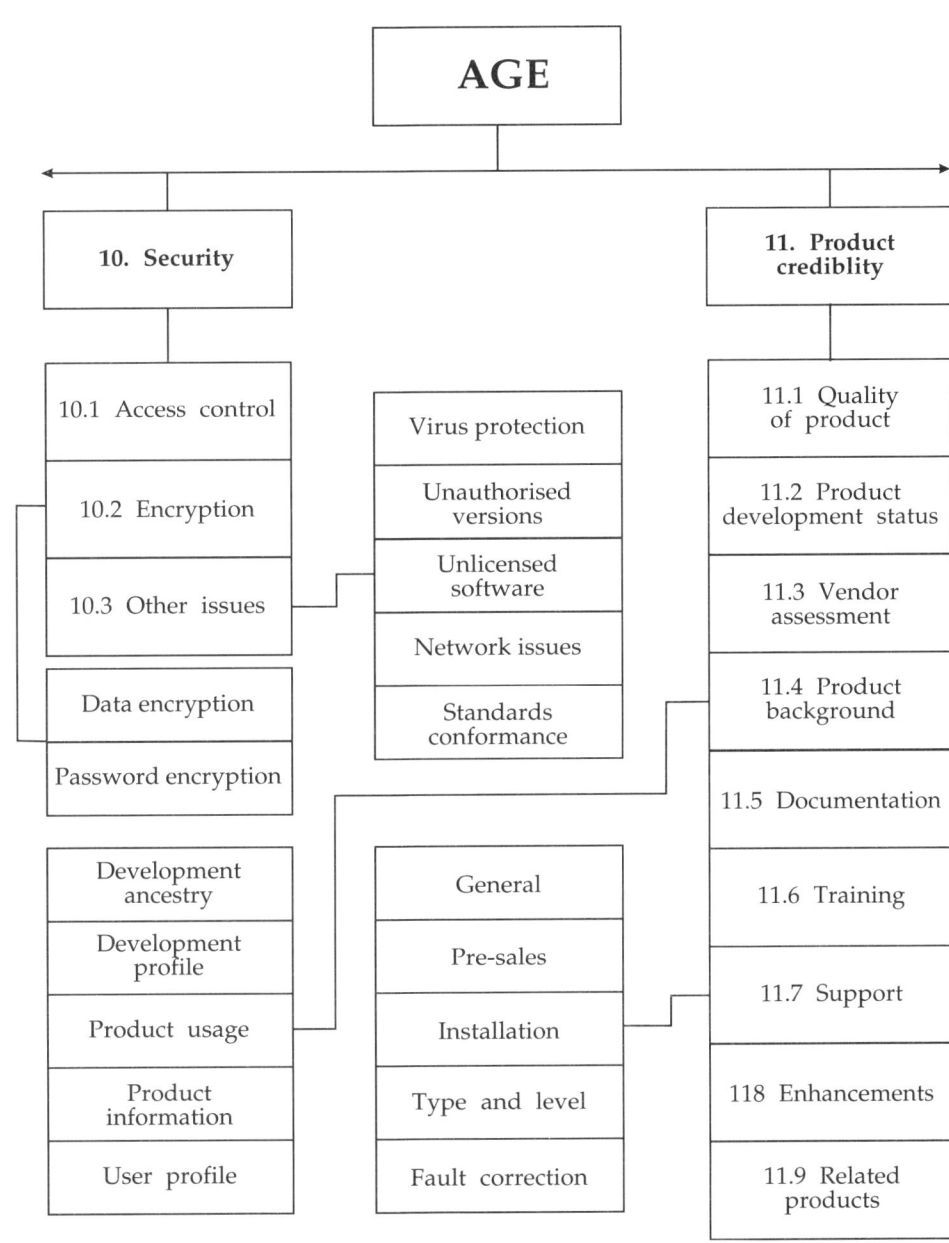

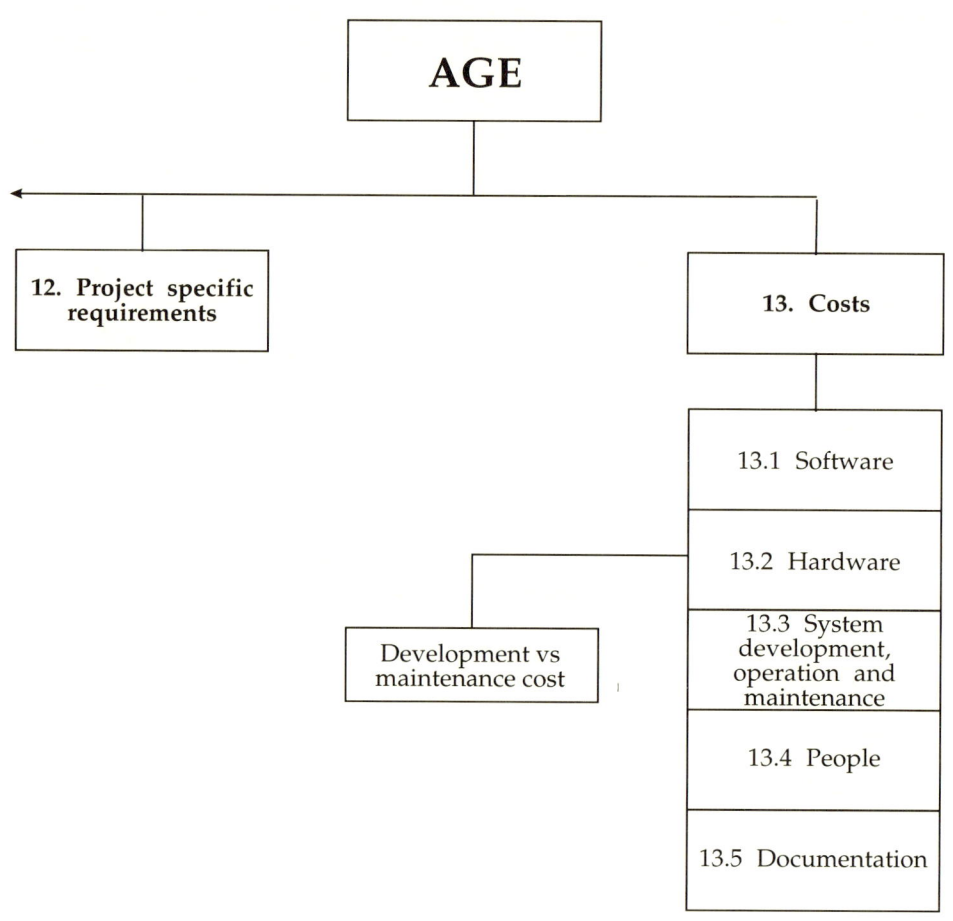

B Bibliography

Appraisal and Evaluation Library

The Appraisal and Evaluation Library is published by CCTA and is available from HMSO Publications Centre. PO Box 276. London SW8 5DT

The following volumes are referenced in this publication:

Overview and Procedures
  ISBN: 0 11 330534 6
CASE Tools
  ISBN: 0 11 330609 1
Database Management Systems
  ISBN: 0 11 330605 9
Knowledge Based Systems
  ISBN 0 11 330570 2
Text-based Information Management Systems
  ISBN 0 11 330 571 0
IT Infrastructure Support Tools
  ISBN: 0 110330586 9

Information Systems Engineering Library

The Information Systems Engineering Library is published by CCTA and is available from HMSO Publications Centre. PO Box 276. London SW8 5DT

The following volumes are referenced in this publication:

A Guide to the SSADM Version 4 Tools Conformance Appraisal Scheme
Testing Criteria for the SSADM Version 4 Tools Conformance Appraisal Scheme

Information Systems Guides

The Information Systems Guides, published by CCTA, are available from John Wiley & Sons Ltd, Baffins Lane, Chichester PO19 1UD.

The following guides are referenced in this publication:

IS Guide B Set: Systems Development Set
ISBN 0 471 92533 0

Appraisal and Evaluation Library
Application Generation Environments Volume

SSADM Documentation   The SSADM Version 4 Reference Manual is published by NCC Blackwell Ltd and is available from NCC Blackwell Ltd, 108 Cowley Road, Oxford, OX4 lJF. ISBN 1 85554 004 5.

PRINCE Documentation   The PRINCE Reference Manual is published by NCC Blackwell Ltd and is available from NCC Blackwell Ltd, 108 Cowley Road, Oxford, OX4 lJF. ISBN: 1 85 554012 6

Other publications   Ovum Evaluates: 4GLs and Client/Server Subscription service
Ovum Ltd

4GLs on UNIX: An Evaluation and Comparison
ButlerBloor

PC Databases: An Evaluation and Comparison
ButlerBloor

Database: An Evaluation and Comparison
ButlerBloor

Annex C
Glossary

C Glossary

4GL	Dictionary based development language which is aimed generally at improving productivity in the building of commercial systems.
application	A suite of one or more related programs that perform a specified task for a user, eg word processing. Application software is contrasted with system software and pure server software, but the distinction is often blurred.
application environment	An integrated set of hardware, communications, software, standards (de jure and other) and methods specific to the execution of application software.
application generator	A system of computer programs designed to facilitate the very rapid implementation of computer systems, especially on-line systems, with minimal recourse to the use of conventional programming.
backup	A utility used to take a copy of the database or subsets of the database usually to tape. This copy can then be used to restore the database in case of serious failure.
batch	A non-interactive process that runs on a queue usually when the system load is at its lowest. The opposite of on-line.
BLOB	Binary Large Object usually applied to unstructured datatypes kept in a database, such as images or unstructured text.
btree	Multilevel index structure optimised for storing and retrieving data in which all leaf nodes are the same distance from the root.
buffer	An area of memory used to receive or send blocks of stored records from or to a storage device eg a disk. A buffer in database terms is a unit of I/O.

cascade-delete	This is a referential integrity operation that allows all foreign key references to be automatically deleted should the primary key be deleted.
CCTA	Government Centre for Information Systems.
client/server	Describes a configuration where an application uses the processing power of both a personal computer and a host system. The personal computer provides an interactive user interface and the host provides large-scale data storage and multi-user information sharing facilities.
commit	Process that completes a logical transaction and then makes the results of the transaction available to other database users.
computer aided software engineering (CASE)	A software component that can be used across the whole development life-cycle of an application system.
concurrency	The ability of a database to handle many users wishing to access a database at the same time.
configuration management	The discipline of identifying components of a system to control changes to it and maintain its integrity throughout its life-cycle.
contention	A problem caused when many users are trying to access the same volatile data. Caused by inadequate locking strategies.
cursor	In SQL this is a pointer to a collection of rows that have been returned by the query that declared the cursor. This allows record at a time processing for application programs.
data definition language (DDL)	Interactive or embedded language for defining the structure (schema) of a database, including table, index and integrity.

Annex C
Glossary

data dictionary	A structured description of a database, it contains the names and structures of all data types, and can also hold information such as processing restrictions, validation rules, and names of programs which access the database.
database	A structured collection of interrelated data that is independent of any applications accessing it. It is created and maintained by a database management system, and a common controlled approach is used to add, delete or modify its contents.
database machine	A computer which is specifically designed to handle databases. These are predominantly parallel in nature, allowing complex operations to be broken down into many small ones which are then executed in parallel.
database management system (DBMS)	Software system designed for creating, updating and retrieving information from a computer database; the software automatically manages the storage and processing of the data comprising the database.
database page	A unit of I/O that could consist of a number of disk blocks.
datatype	A type of data either logical or physical. For example integer, float, character (physical) or money, date and numeric (logical).
DDBMS	Distributed Database Management System.
DDL	Data Definition Language.
DML	Data Manipulation Language.
deadlock	A state in which two or more users are each waiting for resources held by the other(s). Also called a deadly embrace.
dictionary	Usually a store of meta data (data about data) and catalogue information, but could also be used to store forms, user details, programs etc.

149

distributed database	A single logical database that is spread across computer systems at multiple locations connected by a network.
domain	The valid set of values for an attribute in a table.
domain integrity	At the simplest level this is the ability to specify a valid datatype for a column. But is usually extended to handle valid ranges of values, lists or even formats.
encryption	The conversion of a signal or magnetically stored data into coded form for security reasons.
entity integrity	A rule that states that no attribute that is part of the primary key can accept a null value.
Foreign Key	An attribute in one table whose values are required to match those of a primary key in another table. They represent references that allow tables to be brought together usually by a database join.
gateway	A piece of software that provides transparent access to either incompatible file systems, or operating systems.
granularity	Locking assumes that the database consists of a number of items eg rows, pages, tables etc. The place where a lock is taken is its granularity, this granularity could be adjustable either up or down depending on lock resources available.
group commit	Should a number of database users commit at around about the same time, the database management system can reduce I/O by bundling all the commits together and write these to the log in one go.
hash key	The column of a table used for hashing that is also the primary key.
hashing	The use of an algorithm or formula to provide fast access on full key storage and retrievals.

Annex C
Glossary

index	An additional data structure stored on disk which is used to make the search for rows more efficient, when the indexing field is used as part of the query.
indexing field	Columns in a table used to build index structures.
integrity	Refers to the accuracy and validity of data that is held in a database. This is achieved by both the normalisation process and other integrity control mechanisms such as constraints, triggers, and support for rules.
interoperability	The ability of a software product to interface with many other products (eg interfacing a 4GL with many DBMS, spreadsheets, windowing systems etc).
interpretive	Where command statements eg SQL are checked, translated and then executed, as opposed to compiled, which just executes.
JCL	Job Control Language.
join	The process where two tables are connected on the basis of common data. (primary and foreign keys).
journalling (before and after image)	Records of portions of the database that are logged before a change or after a change has been made. These records can then be used to recover a database in case of failure.
key (primary)	An attribute or data item that uniquely identifies a row in a table.
link structure	Data structure that uses pointers to connect records in a logical sequence, also called a pointer chain.
local	A system to which the user is directly connected.
log	A record of all changes that have taken place on a database. (See also Journalling) .

multithreaded	A process that can support a number of users but only has one copy of the object code in memory. This makes for more efficient use of CPU because it diminishes problems of context switching. It also reduces memory requirements.
network DBMS	A DBMS where the relationships between record types (or entities) is stored in the database forming a network of entities and relationships.
non-procedural language	A language where the order of statements is not important, and where there is an emphasis on what is to be done (as opposed to how it should be carried out).
object	The instance of a class of objects, or object type. Objects can be thought of as records with the functions that might apply to them (eg update customer balance in a CUSTOMER record).
object DBMS	A DBMS which supports the storage and retrieval of objects. Such products normally provide interfaces to OO languages.
object orientation	A set of concepts embodied which promote the association of functionality with data structures. This concepts are embodied in languages such as Smalltalk and C++.
object oriented programming	The concept of procedure and data, embodied in conventional programming systems, is replaced by the concepts of objects and messages. An object is a package of information and a description of its manipulation, and a message is a specification of one of an object's manipulations.
off-line function	A function where all the data is input and the whole of the database processing for the function is completed without further interaction with the user.

Annex C
Glossary

OLTP	On-line Transaction Processing is typified by having frequent row updates or insertions to a database, requiring fast response.
on-line function	A function where the system and the user communicate through input and output messages, ie message pairs. The system responds in time to influence the next input message. On-line functions may include off-line elements such as printing an off-line report.
optimiser	The component of a database management system that determines the best way to satisfy a query.
optimistic locking	An assumption that in the majority of transactions, contention for a given resource (Row, Page etc) will not occur. Data is not locked, but transactions are rolled back if any inconsistency occurs.
portability	The ability to move an application environment from one hardware platform to another.
precompiler (SQL)	A piece of software which converts SQL statements into a native data manipulation language that an ordinary compiler can recognise.
preprocessing (SQL)	The act of submitting a program that contains SQL statements to the pre-compiler.
Primary Key	An attribute or data item that uniquely identifies a row in a table.
procedural language	A language where the order of statements is important, and where there is an emphasis on how things are to be done (as opposed to what).
process	A unit of a program that is being executed by the CPU. A program could generate more than one process.
production systems	Systems which deal with the day to day running of a company (eg accounts, purchase ordering etc).

QBF	Query By Forms.
query based system (QBS)	A system that mainly consists of read only access.
RDBMS	Relational Database Management System.
real-time	Used to describe processing where a computer system accepts and updates data at the same time feeding back immediate results that influence the data source.
referential integrity	An integrity constraint which specifies the value or existence of a foreign key is dependant upon the value or existence of a primary key.
relational database	A database that is a collection of two-dimensional tables.
replication	A copy of data used in a distributed system to provide local rather than remote access to frequently used data.
report writer	A piece of software used to quickly generate formatted reports.
repository	A dictionary normally associated with the storage of objects created by CASE products.
requester server	See client/server.
rollback	A recovery technique that allows a database to be restored to an earlier state by undoing a transaction.
row	A row is similar to a record in a file, consisting of a collection of fields.
runtime	When a program executes.
select (SQL)	In relational algebra a command that extracts selected rows of a table according to a search criteria.
server	A process that provides a service, usually to the database.

Annex C
Glossary

third party vendor	A Software vendor that does not sell hardware.
three schema architecture	A database architecture that provides three levels or views of data. The conceptual level (the overall logical model), the external level (user or applications view of the database), and the internal level (the description of the physical database structure). This separation protects each level from changes to the other levels.
throughput	The rate at which transactions are completed against a database system.
timestamp	The adding of date and time information to a database page or database row for purposes of consistency and concurrency.
tokenised code	An intermediate form of machine independent code which is interpreted at runtime.
trigger	This specifies an action that must be taken should a particular condition occur.
two phase commit	An approach to the commit process in distributed database systems in which there are two phases. In the first phase each participating node is instructed to prepare to commit and must respond as to whether the commit is possible. After each node has responded correctly then the second phase consisting of the actual commit can continue otherwise the transaction must be aborted and rolled back.
view	This is a virtual table that is not physically stored anywhere in the database.
virtual field	These can be derived from actual column values, but are not physically stored anywhere. Typical examples of such fields are totals.